HERITAGE

BATTLEFIELDS
Of NORTH AMERICA

HERITAGE

BATTLEFIELDS
OF NORTH AMERICA

IAN WESTWELL

COMPENDIUM

This edition published in 2008 by

COMPENDIUM

ISBN: 978-1-906347-56-7

Copyright © 2008 Compendium Publishing Ltd
43 Frith Street
London W1D 4SA
United Kingdom

Cataloging-in-publication data is available from the Library of Congress.

The Author
Ian Westwell is a military historian and author who has written and /or contributed
to several military books and periodicals. Among his most recent works have been
*World War One Day By Day, The Encyclopedia of World War I, In the Path of the Third
Reich,* and *1st Infantry Division.*

Project Manager: Ray Bonds
Designer: Cara Rogers
Color reproduction: Anorax Imaging Ltd

Printed and bound in China

Page 1: An unknown artist's depicts the
union advance toward Dunker Church
during the battle of Antietam.

Pages 2–3: A 1781 hand-colored French
etching showing the British surrender at
Yorktown—rather oddly shown as a
walled medieval town. The French, in
blue, are in the foreground; the
Americans, in red, are in the background.
Library of Congress ar301100

Page 4: The Battle of New Orleans 1815
after E. Percy Moran, c. 1910. Percy
Moran was a specialist in colonial and his-
torical subjects; his uncle was the famous
artist Thomas Moran. *Library of Congress
3f03796*

CONTENTS

INTRODUCTION

The people of the continental United States have to all intents and purposes not seen the face of conventional battle since the ending of the prolonged Native American Wars, those intermittent conflicts that reached their largely inevitable conclusion with the events at Wounded Knee on 29 December 1890—the year in which, coincidentally, the US Census Bureau announced that there was no longer a line of frontier settlement in the Mid-West. Yet for the 300 or so previous years the developing United States had been ravaged by conflicts in which European colonists fought European colonists, in which Native Americans fought Native Americans and the colonists, in which nascent American citizens fought each other or their English overlords, and—most devastating of all—Americans struggled against each other.

WAR IN COLONIAL AMERICA

The Americas were first settled by European colonialists in the early 17th Century and there was almost immediate conflict between the various incomers, chiefly English, Dutch, French, and Spanish, as well as between the new settlers and the indigenous Native Americans. In the case of the colonialists, the conflicts were often an extension of European-based dynastic conflicts. For example, Queen Anne's War (1702–1713),

in which not for the first time England and France clashed, was an extension of the War of the Spanish Succession in Europe, while King George's War (1740–1748), which involved the same protagonists, was an off-shoot of the War of the Austrian Succession.

Native Americans acted as allies in these wars. During Anglo-French King William's War (1689–1697), itself part of the War of the League of Augsburg that pitched France against England, Germany, and Spain, the Iroquois fought alongside the English while various other tribal groupings sided with the French. On other occasions the colonialists used animosities between Native American tribes to their own advantage. The Dutch, who were seeking to grab a greater part of the fur trade, provided the Iroquois with arms to fight the French and their Native American allies, thus provoking a war that began in 1642 and carried on for the next eleven years. Yet on many more occasions Native Americans and the colonialists battled against each other. The first European colony, that established by the English at Jamestown in 1607, was the site of a massacre at the hands of Native Americans in 1622, an event that set the tone for the relationship between the two groups for the next 250 years or more.

RIGHT: A clash between colonists and Native Americans during King Philip's War (1675–77), the first significant military conflict in the American colonies, which cost the lives of some thousands. King Philip was the supreme chief of the Wampanoags, one of several tribes who joined the uprising against the imposition of European authority across New England.

FAR RIGHT: A map showing some of the very many major military clashes that have taken place across the United States, from Bunker Hill (1775) during the War of Independence (1775–83) to the attack on Pearl Harbor (1941) that heralded the country's entry into World War II (1941–45).

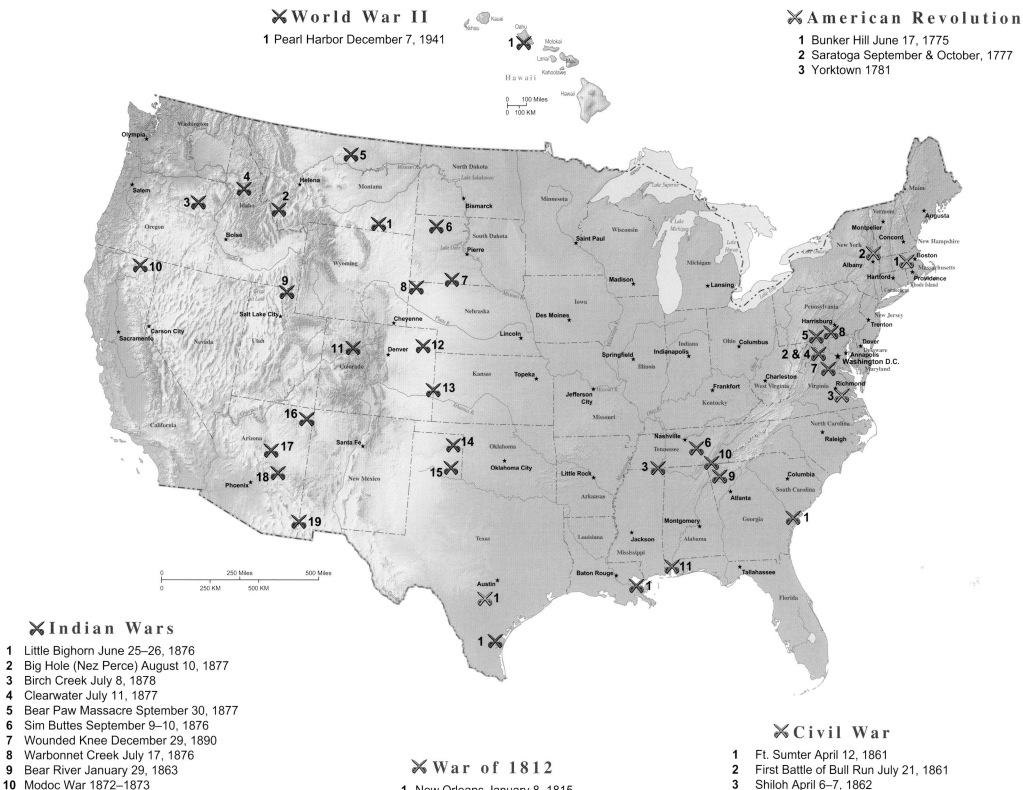

⚔ World War II

1 Pearl Harbor December 7, 1941

⚔ American Revolution

1 Bunker Hill June 17, 1775
2 Saratoga September & October, 1777
3 Yorktown 1781

⚔ Indian Wars

1 Little Bighorn June 25–26, 1876
2 Big Hole (Nez Perce) August 10, 1877
3 Birch Creek July 8, 1878
4 Clearwater July 11, 1877
5 Bear Paw Massacre Sptember 30, 1877
6 Sim Buttes September 9–10, 1876
7 Wounded Knee December 29, 1890
8 Warbonnet Creek July 17, 1876
9 Bear River January 29, 1863
10 Modoc War 1872–1873
11 Meeker Massacre September 17, 1879
12 Summitt Springs July 11, 1869
13 Sand Creek Massacre November 29, 1864
14 Washita November 27, 1868
15 Palo Duro Canyon September 28, 1874
16 Canyon de Chelly January 12, 1864
17 Big Dry Wash July 17, 1882
18 Salt River Canyon December 28, 1872
19 Apache Pass July 15–16, 1862

⚔ War of 1812

1 New Orleans January 8, 1815

⚔ Texan Revolution

1 Alamo February & March, 1836

⚔ Mexican–American War

1 Palo Alto May 8, 1846

⚔ Civil War

1 Ft. Sumter April 12, 1861
2 First Battle of Bull Run July 21, 1861
3 Shiloh April 6–7, 1862
4 Second Battle of Bull Run August 29–30, 1862
5 Antietam Sepember 17, 1862
6 Stones River December 31, 1862
7 Chancellorsville May 1–4, 1863
8 Gettysburg July 1–3, 1863
9 Chickamauga September 18–20, 1863
10 Chattanooga (inc. Battle of Lookout Mountain & Battle of Missionary Ridge) November 23–25, 1863
11 Mobile Bay August 5, 1864

BELOW: The Battle of Monongahela, fought a little to the south of what is now Pittsburgh on 9 July 1755, was one of the worst defeats suffered by the British during the French and Indian War (1754-63). Ambushed by a mixed French, French-Canadian, and Native America force, the British and American column suffered 1,000 casualties out of a total of some 1,300 men. The ambushers recorded just forty-three killed or wounded.

RIGHT: US minutemen, part of a militia unit commanded by Captain John Parker, return fire on the light companies of the British 4th and 10th Regiments of Foot under Major John Pitcairn at Lexington Common, 19 April 1775. This clash, the first of the War of Independence, left eight of Parker's men dead and ten wounded, while one of Pitcairn's sergeants was slightly wounded. The British would suffer a total of 73 killed, 174 wounded, and 26 missing presumed captured as they retreated back to Boston.

Estimates suggest that there were something in the region of sixteen million Native Americans scattered across the Americas, both North and South, at the time of the opening phase of European colonization. Yet despite their obvious numerical advantages, they singularly failed to halt the expansion of the various colonial enclaves. There was no single reason for this. The Native Americans were, at least initially, better adapted tactically to fight in the local terrain, and their weapons were no less effective than those of their opponents. Their most obvious weaknesses appear now to have been their lack of political cohesiveness—tribes that had often been at odds with each other made at best rare and uneasy allies—and their lack of battlefield discipline. Nevertheless, the Native Americans caused the European settlers many problems, chiefly due to their frequent raids along the various frontier zones. Gradually, the conflicting interests in North America were whittled down—largely by the end of the 17th Century, until only those involving the English, French, and Native Americans remained.

THE BIRTH OF THE UNITED STATES

Europe's colonial rivalries in North America were effectively ended by the so-called French and Indian War (1754–63), which was mostly part of the wider European-based Seven Year's War. English hegemony over North America was, despite a number of setbacks, finally secured in September 1759 with victory at the Battle of the Plains of Abraham outside Quebec. Yet England's control of its new North American empire was far less secure than it at first seemed and would last for less than a mere three

ABOVE: General George Washington led his battered command across the Delaware River into Pennsylvania in December 1775, after he had suffered a number of defeats around New York over the previous few months. The British failed to follow his battered army, preferring to go into their winter quarters, and Washington soon struck back. He re-crossed the river with 2,400 troops on the 25th, the event depicted here, and then defeated a force of 1,400 Hessian troops fighting for the British at Trenton, New Jersey, the next day.

RIGHT: Shawnee Indians led by Tecumseh rebelled against the US authorities in 1811. A force of a thousand troops under the governor of the Indiana Territory, William Henry Harrison, was sent against the Shawnee capital, Prophetstown, 150 miles north of Vincennes. Harrison's camp on the Tippecanoe Creek was attacked by around 700 Native Americans at dawn on 8 November and he only narrowly avoided being defeated, losing 37 dead and 150-plus wounded. Here, Harrison, who earned the nickname "Old Tippecanoe," leads a counterattack mounted on his gray charger. In reality, most of the US troops, cavalry included, fought on foot.

arguably the world's greatest maritime nation, one with a largely unmatched navy, and had few military resources of their own—no standing army, no navy, and, at least at first, no allies to support them. It was perhaps inevitable that the early phase of the war went badly for the revolutionaries. Admittedly the British won a largely pyrrhic victory at the Battle of Bunker Hill outside Boston in June 1775, and were forced out of the New England port the following March, but there was little else to cheer the independence movement. An attack on Canada in late 1775–1776 was a fiasco, and a campaign around New York and across New Jersey between July and December 1776 appeared equally without merit.

Yet, even at this most dark moment, the war was turning against the British although it had several more years to run. The Declaration of Independence announced in July 1776 was a statement of intent that rallied waverers to the cause, but equally important was the announcement of tangible support from both France and then Spain. There was, at the very end of 1776, better news from the battlefield— first came the victories at Trenton and Princeton in late 1776 and early 1777 but, probably of greater significance, was the crushing victory at Saratoga in October 1777, in which a British field army was forced into ignominious surrender.

The revolution was far from secure, however, and the winter of 1777–1778 that the main revolutionary army spent miserably encamped at Valley Forge marked a nadir in its fortunes. On the positive side, France declared war on Britain in June 1778 and Spain followed suit twelve months later. The tide really turned in 1780, a year in which a force of Loyalists was roundly defeated at the Battle of King's Mountain, revolutionary general Nathanael Greene ran rings around the British in the south, and French troops arrived to tangibly back the revolution. The British in North America, overstretched and struggling to maintain a viable supply line across the North Atlantic, finally succumbed in October 1781, when their main field army was forced to surrender at Yorktown after a siege of some three weeks. A new nation, the United

decades. The causes of the American War of Independence that erupted in April 1775 were various—part constitutional, part ideological, part demographic—but the conflict pitched the existing thirteen colonies against their imperial masters as well as turning pro-English "Loyalists" or "Tories" against their fellow but pro-revolution brethren. Just for good measure, Native Americans were also drawn into the conflict.

At first sight the scales weighed heavily against the revolutionaries. They faced

ABOVE: A panoramic and accurate view of the Battle of New Orleans on 8 January 1815, looking eastward from the west bank of the Mississippi river and showing the US defensive position, Line Jackson (left), being assaulted by the British.

RIGHT: Major-General William Henry Harrison, on the gray horse, shoots down the Shawnee chief Tecumseh during the Battle of the Thames, near London, Ontario, on 5 October 1813, an episode in the War of 1812 (1812–15).

States, was finally born with the signing of the Treaty of Paris in November 1782, and ratified by Congress in April the following year.

PRESERVING AND EXPANDING THE NEW NATION

Independence did not bring prolonged peace to the new nation, and the next century or so saw renewed conflicts with old foes and former allies as well as a battle for the very soul of the United States. There were brief periods of internal unrest, such as those uprisings known as Shay's (1786–1787) and the Whisky (1794) Rebellions, but these were relatively minor affairs when compared to conflicts with the Native Americans and British, ones fought largely to expand and defend, respectively, the borders of the United States. The first major post-revolutionary event was war with the Shawnee in 1811, whose leader, Tecumseh, was backed by Anglo-Canadian fur interests. The US victory at the Battle of Tippecanoe in November saw the Shawnee defeated if not entirely vanquished, and also led to an outpouring of anti-British sentiment. A mere seven months later the two countries were at war.

The War of 1812, which actually lasted until 1815, in many ways echoed the War of Independence. The British mostly, if not entirely, bested the part-militia, part-regular US forces in a number of campaigns along the US–Canadian border and went on to burn Washington, D.C., during August 1814 in retaliation for the US destruction of York (modern Toronto), an event in April 1813 that is now believed to have been provoked by the accidental explosion of a powder magazine that killed or wounded some 320 US troops. The British, who were heavily involved in the ongoing wars against Napoleon, were never able to effectively fight two separate wars on two widely dispersed fronts, but made one last effort to secure some sort of victory. The Battle of New Orleans, which was fought in January 1815, was an unmitigated disaster for the British and, absurdly, took place after a peace settlement, the Treaty of Ghent, had been signed the previous month.

Once again, Native Americans had been drawn into the conflict. Tecumseh, for example had fought and died at the Battle of the Thames in October 1813, while the Creeks in Alabama sided with Britain only to be roundly defeated at the Battle of Horseshoe Bend in March 1814. The War of 1812 effectively secured the future of the United States, and its founding fathers now turned their attention to expansion. Some lands were bought. The 1803 Louisiana Purchase of land from France had already added the watershed of the Mississippi and Missouri Rivers, doubling the nation's size at a stroke, while the buying of Alaska from Russia on 30 March 1867 would add a further sizeable piece of territory.

However, much of this territorial expansion was achieved at the point of a bayonet. Various Native American tribes initially bore the brunt of the push out from the original colonies by the white settlers. There were several wars against the Seminoles of Florida, beginning in 1818 and concluding only in 1843, as well as the short Black Hawk War of 1832. Conflict was not confined to white settlers and Native Americans, and brought the United States into conflict with colonial Spain. Warfare erupted in June 1835 when American settlers in Spanish-controlled Texas revolted, an event that was confirmed by the settlers' declaration of independence on 2 March 1836, Although the rebels were defeated during a brief if momentous siege at a former mission, The Alamo, four days later, independence was secured with a one-sided victory at the Battle of San Jacinto the following month, and the United States gained its twenty-eighth state in 1845.

While the Spanish could stomach, if reluctantly, the emergence of the Texan republic, they made vocal their opposition to Texas joining the United States. Equally, the latter was in an expansionist mood as part of its Manifest Destiny to "spread itself across the whole continent" and, almost inevitably, the two sentiments proved irreconcilable. War broke out in March 1846, although a formal US declaration came only in May. There were three main areas of operation—southern Texas and northern Mexico, the central region of Mexico, and California. The Battle of Palo Alto on 8 May, five days before the formal outbreak of hostilities, helped secure Texas from

ABOVE LEFT: US troops begin their attack on Chapultepec on 13 September 1847, the final major battle of the US-Mexican War (1846-48).

LEFT: After defeating Mexican troops at the Battles of Palo Alto and Resaca de la Palma in 1846, Major-General Zachary Taylor led 6,000 men of his Army of the Rio Grande into northeast Mexico on 18 May, capturing Monterrey on 24 September and winning the tightly contested Battle of Buena Vista on 22–23 February 1847. The latter victory effectively ended the northern campaign.

invasion and paved the way for the US occupation of Monterrey in September.

Mexican resistance in California was effectively crushed at the Battle of San Gabriel in January 1847, while the crowning moment of the war came the following September, when US forces took Mexico City. Peace negotiations reached fruition with the ratification of the Treaty of Guadalupe Hidalgo in July 1848, which confirmed the border of Texas along the Rio Grande and added another large slice of territory to the United States, including what are now California, Nevada, Utah, most of Arizona, New Mexico, and parts of Colorado and Wyoming.

WAR BETWEEN THE STATES

Over the next two decades or so the process of expansion and consolidation across the continent continued and brought the relatively small US Army once again into conflict with various Native American tribes. There were expeditions rather than out-and-out conflicts, mostly small-scale raids and counter-raids, and there were some thirty "small wars" recorded in the fifteen years after 1850. Yet, even as the nation pushed westward, it was struck by the most traumatic episode in its comparatively short history, various described as the American Civil War and the War Between the

ABOVE: Commander of the Veracruz garrison, General Juan Morales (or one of his officers), surrenders to Major-General Winfield Scott after the Mexican city had been pounded by artillery fire during 22–27 March 1847.

LEFT: During the Civil War, President Abraham Lincoln was looking for a war leader who would be able to defeat the determined Confederate forces in the field of battle. In 1864 he found one in Ulysses S. Grant, seen here at far left on Lookout Mountain, near the battlefield of Chattanooga. Tough and resolute, the wily Grant used finesse, speed, and generally sound tactics to grind down the Confederate forces led by the equally tenacious Robert E. Lee. Another attribute they shared was respect for their respective enemies. Cometh the hour, cometh the man....

RIGHT: When General Ambrose Burnside sought to march on Richmond, Virginia, the Confederate capital, he was roundly defeated at the Battle of Fredericksburg (seen here) on 13 December 1862.

States. Fought between the northern Union and the southern Confederacy, it had many causes but revolved around who had preeminence—the individual state or the greater Union. Whatever the background, it pitched the smaller, agricultural, slave-owning Confederacy against the larger, more industrialized North.

On paper at least, the war should have been short-lived. The North, which was most likely to triumph in a prolonged conflict, had the greater resources in virtually every sense, yet the bloody conflict was to last from April 1861, when Union-held Fort Sumter was bombarded by Southern artillery, to May 1865, when the last Southern troops in the field laid down their arms. However, the South did have some advantages in the first years of the war—its troops tended to perform better, its commanders were the more able, and many battles were fought on home (Southern) terrain. The war was fought largely by citizen-soldiers, since the peacetime US Army was tiny and the new soldiers had to adapt their fighting modes to reflect various advances in military technology that vastly increased the range and lethality of weapons.

The war began with rigid linear formations and even cavalry charges but would gradually evolve into one in which more flexible battlefield formations, the use of natural cover, and siege warfare would become increasingly prevalent. Nevertheless, the war remains the most sanguine in United States history, killing some 718,000 men and wounding at least a further 375,000. The bloodiest single-day battle came early—on 17 December 1862—when almost 23,000 Union and Confederate troops became casualties (killed and wounded) during the Battle of Antietam.

the South ever taking Washington. Equally significant, if less lauded, was the North's capture of Vicksburg on 4 July, an event that split the South in two.

Pressure from the North steadily mounted as the war entered its final phase, especially when the Union's naval blockade of the South, the Anaconda Plan, really began to bite. The war became less and less one of maneuver and was increasingly about attrition—the North simply began to grind down the South's ability to wage war. The latter's field armies were slowly battered into submission in various bloody encounters, while its limited industrial base and agricultural heart were ravaged, not least by General Sherman's 300-mile "March to the Sea" from Atlanta to Savannah in late 1864. The South's most renowned force, the Army of Northern Virginia, capitulated on 9 April 1865.

THE END OF NATIVE AMERICAN RESISTANCE

The process of reconstruction, of reuniting the war-torn states, would take many years, but the process of expansion westward hardly paused—between 1865 and 1898 the US Army fought close to a thousand actions in a dozen separate campaigns against various Native American tribes. Although the tide was running against the latter, the various wars and campaigns did not see the technologically superior US Army having things all its own way. The Native Americans increasingly shied away from direct battle and increasingly turned to a guerrilla type of warfare in which hit-and-run raids predominated. The Native Americans could also draw on some naturally talented war leaders, men like Crazy Horse of the Sioux, who destroyed General George Armstrong's 7th Cavalry Regiment at the Battle of the Little Bighorn in 1876, and Chief Joseph of the Nez Percé, who the next year led the Army on a near 2,000-mile merry dance across Oregon and the Idaho, Montana, and Wyoming Territories before being forced into surrender when confronted by a US expeditionary force ten times the size of his tribe.

The land war was largely fought in two main theaters, effectively the area between the rival capitals, Washington and Richmond, which lay a mere 100 miles or so apart, and along the Mississippi River. For the first years of the war in the former theater, both sides attempted to capture the other's capital but were largely frustrated, although in their invasions of the North the Confederates had the upper hand on the battlefield. It was not until 1863 that the war can be said to have swung decisively in favor of the Union—the Battle of Gettysburg, fought over1-3 July, ended all hope of

ABOVE: The final act of the prolonged wars against the Native Americans was the Battle of Wounded Knee on 20 December 1890.

RIGHT: The Apache leader Geronimo (far right), photographed with some of his warriors. The chief was a cunning and ruthless leader, one who waged a form of partisan warfare in the southwest United States that largely bamboozled the US Army in the 1880s. He surrendered in both 1882 and 1885 but on each occasion went back to his old ways, finally surrendering for good at Skeleton Canyon, Arizona, in September 1886.

FAR RIGHT: Although conventional warfare has not touched the United States since the Japanese attack on Pearl Harbor, the nation has not been immune to other forms of aggression. On 11 September 2001 members of the Islamic group Al-Qaeda hijacked four passenger aircraft. One, the Los Angeles-bound American Airlines Flight 77, was crashed into the Pentagon in Washington, D.C., killing sixty-four passengers, crew, and terrorists as well as 125 people on the ground.

The last serious campaign was that waged by Apache leader Geronimo in Arizona, New Mexico, and Mexico between 1885 and 1886, a guerrilla campaign that ended with his surrender. The final acts of the prolonged Native American Wars came in 1890. Sitting Bull was killed in a skirmish in South Dakota on 15 December, and five days later many of his followers were killed in the fighting at Wounded Knee by troops from the reconstituted 7th Cavalry. The wars between the various inhabitants of the United States were over, but the country itself, although inwardly peaceful was not entirely immune from external conventional military attack.

US troops were involved in a number of overseas operations over the following decades, notably the Spanish-American War of 1898, and also fought in World War I in Europe, but their homeland remained unmolested. However, all that changed on 7 December 1941, when the Japanese launched a surprise attack on Pearl Harbor, the Hawaiian home of the US Pacific Fleet. The attack was a body blow, one that pushed the country into World War II, but the damage was not as severe as first thought and within six months the US Navy had inflicted a devastating blow on its Japanese counterpart at the Battle of Midway. Since 1941 the US has fought its wars overseas, although, as the events of "9/11" have shown, it is far from immune from unconventional attack.

THE BATTLE OF BUNKER HILL, 1775

DATE: 17 June 1775

COMMANDERS: (American) Major-General Artemas Ward; (British) Major-General William Howe

TROOP STRENGTHS: (American) 4,000+; (British) 3,500

CASUALTIES: (American) 134 dead, 379 wounded and 30 captured; (British) 226 dead and 900+ wounded

The American War of Independence was but two months old when on the night of 16-17 June 1775 some 1,200 soldiers from Massachusetts under Colonels William Prescott and Richard Gridley began to mistakenly fortify Breed's Hill on the Charlestown Peninsula. They had, in fact, been ordered by Ward to occupy Bunker's Hill, which lay to its rear and nearer to the peninsula's narrow neck. Breed's Hill's high ground was nevertheless a position from where Ward's forces overlooked and could dominate the British-held port of Boston, which had been under siege since late April. During daylight the Massachusetts men, who had constructed a redoubt, breastwork, and three fleches (open-ended, vee-shaped earthworks) in double-quick time, were joined by further troops from Connecticut and New Hampshire, the latter led by Colonel John Stark.

The British response came at 0400 hours, when one of their warships opened fire, soon followed by others. The commander of the Boston garrison, Lieutenant-General Thomas Gage, ordered Howe to launch an amphibious assault on the peninsula but,

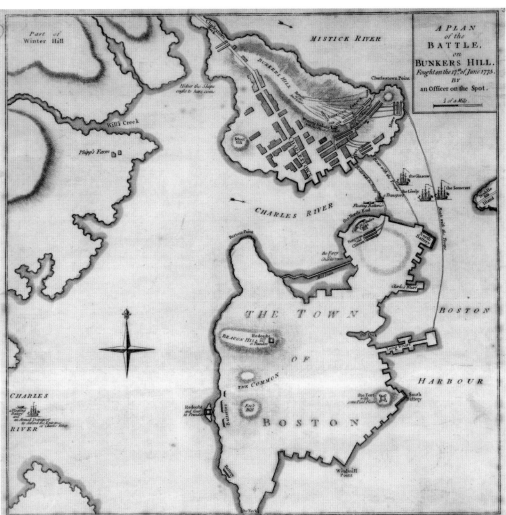

LEFT: British warships bombard the American earthworks on Breed's Hill on 17 June. Further reinforcements arrive by boat at Moulton's Point on the Charlestown Peninsula to bolster their hard-pressed comrades slogging their way up the hill in the face of murderous fire from the position's defenders

ABOVE: A reasonably accurate map of Boston and its environs that shows the various routes taken by the British to reach the Charlestown Peninsula and the points where they landed. The first troops came ashore at Moulton's Point at the east tip of the peninsula, but secondary landings also took place outside Charlestown.

RIGHT: Various maps depicting parts of New England around Boston, the town itself, and the Charlestown Peninsula and the port's harbor. The actual fighting began with the British expedition to Lexington and Concord on 19 April 1775, following which Boston was placed under siege.

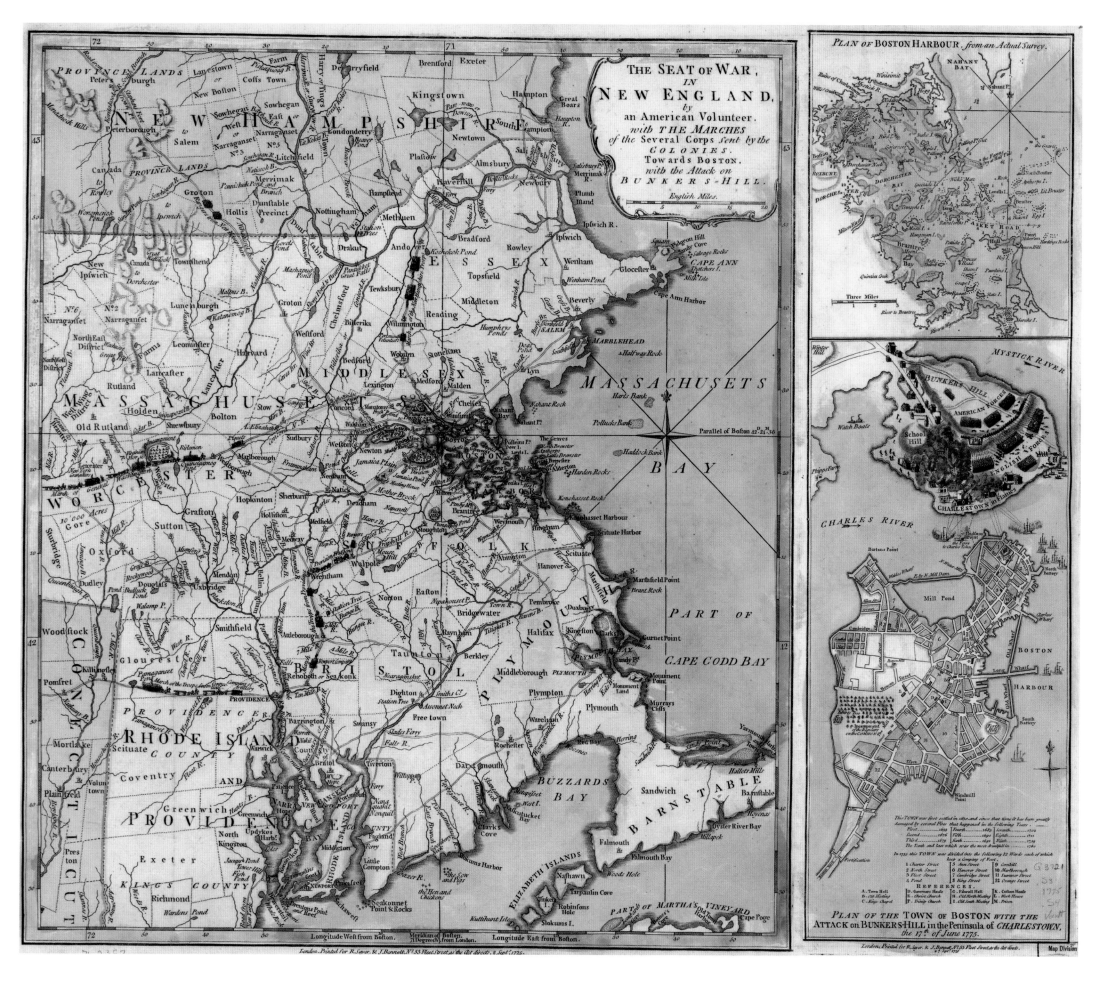

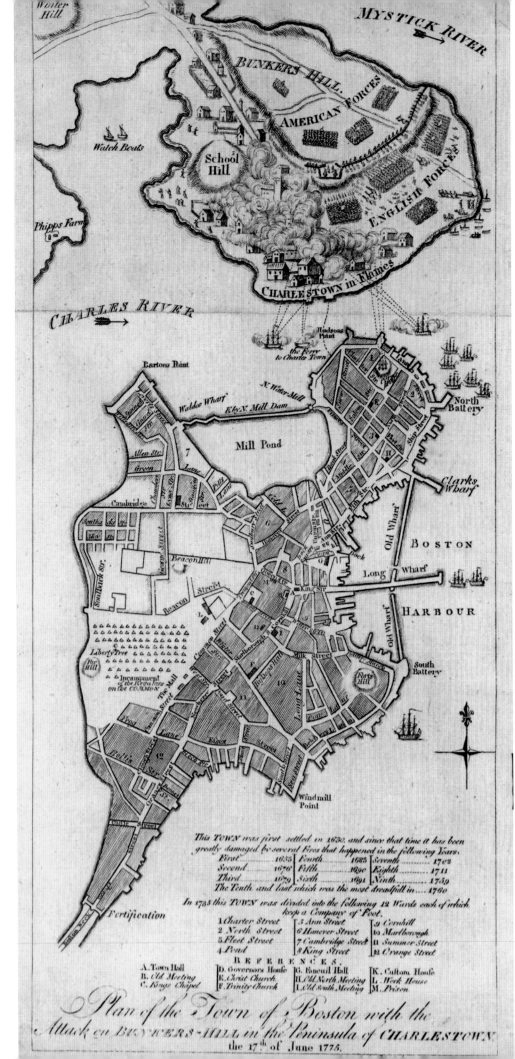

LEFT: A map of the battle showing the British making one of their several attacks up Breed's Hill and the American defenses stretching from the peninsula's shoreline on the Mystic River to the main redoubt on the hill's summit. It also depicts several British warships setting fire to Charlestown, although only one, the frigate HMS *Lively*, was responsible. This took place at around 1600 hours to force American sharpshooters, who had been firing on the British left flank, out of the buildings prior to the beginning of the final, successful British push.

RIGHT: Dr. Joseph Warren was a leading light in the revolutionary movement and, as a major-general, fought at the Battle of Bunker Hill. He was killed instantly by musket fire as the first wave of British troops—Royal Marines and the 47th Regiment of Foot—stormed the redoubt on 17 June.

RIGHT: Although this map gives a general feel of the battle, it contains certain inaccuracies. Bunker's Hill is far too elongated; in reality it occupies only the forward central part of the peninsula and does not stretch down to the shores of the Mystic River. Charlestown is also too large.

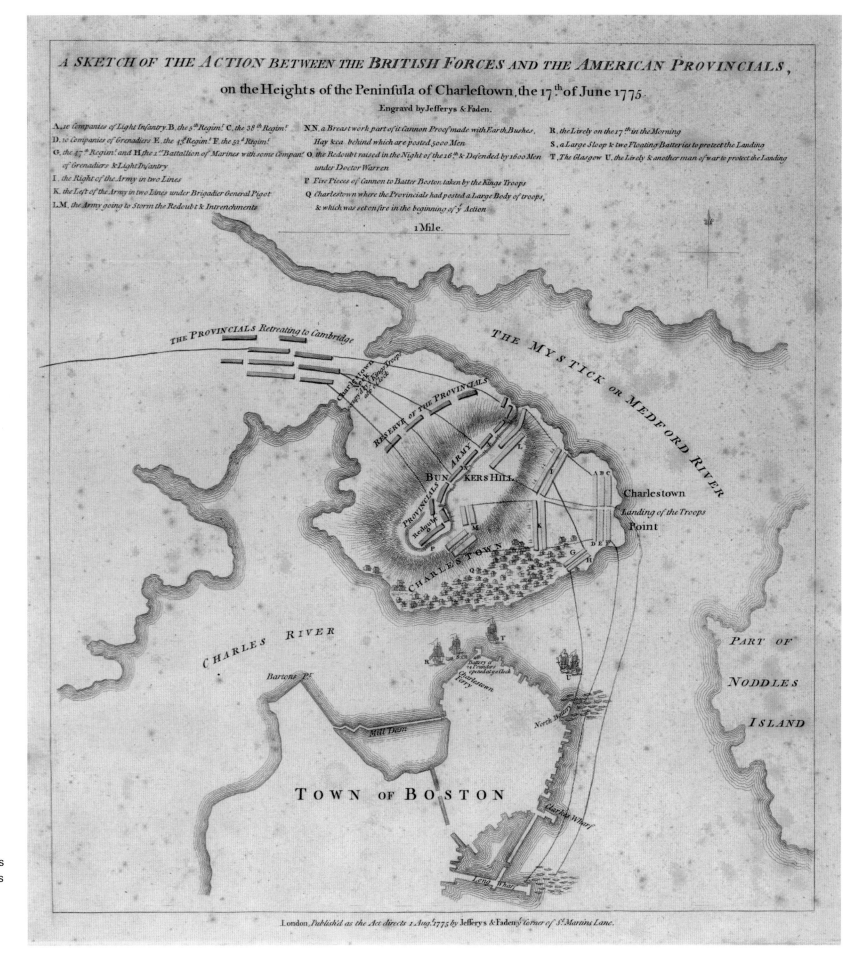

LEFT: Lieutenant-General Thomas Gage was the British commander-in-chief in North America and also governor of Massachusetts but was largely ineffective in both roles. His general lack of decisiveness led those under him to nickname him "Old Tom" or "Granny Gage."

because of a shortage of rowing boats, his force would have to be landed in two waves. The first of these came ashore at Moulton's Point at around 1400 hours and the second some sixty minutes later. With all of his men at hand, Howe now opted to feint against the redoubt and make his main thrust to the northeast along the banks of the Mystic River. However, he had underestimated the fighting abilities of his opponents, and Stark's men, who were positioned behind a rail fence by the Mystic, stopped his advance in its tracks.

The British commander now decided to concentrate his efforts against the redoubt. Three times his troops attempted to storm the position at the point of their bayonets and on the first two occasions they were cut down in droves by the accurate fire of Prescott's men, but the Americans were running increasingly and worryingly low on ammunition. The third attack finally did the trick, with the British breaking into the redoubt at around 1630 hours. They harried but were unable to destroy their retreating opponents and, wholly exhausted, ended the battle when they reached the neck of the peninsula around an hour later. Major-General Henry Clinton arrived at 1800 hours and officially called off the attack.

The British had won the battle but at a terrible cost, with roughly 40 percent of their troops becoming casualties, and they were profoundly shocked by the way that their opponents, whom they hardly regarded as "proper" soldiers, had stood up to them. Their costly victory did not bring about the end of the siege of Boston. That would continue under the direction of General George Washington, and the British were destined to abandon the city for good on 17 March 1776.

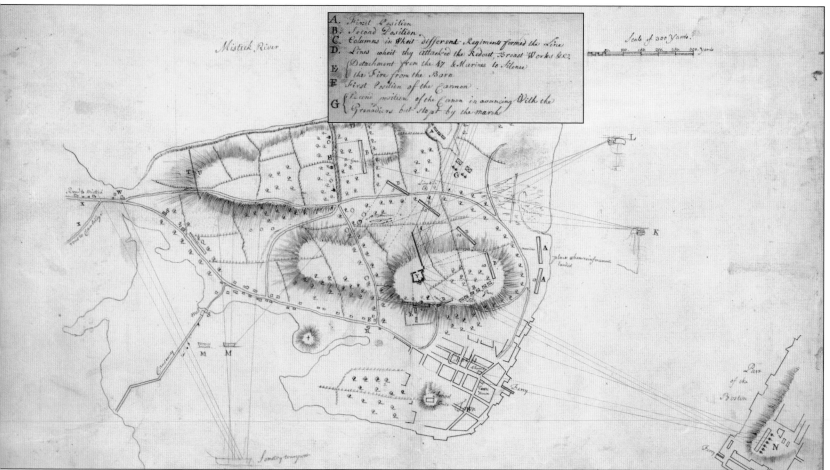

ABOVE: British grenadiers attacking up Breed's Hill in the final stages of the battle. This was a composite battalion of men drawn from the grenadier companies of several regiments. These men have blue facings on their coats and are therefore from either the 4th, 23rd or amalgamated 18th/65th Regiments of Foot. In the battle, the battalion actually attacked part of the breastwork running out from the redoubt.

LEFT: A largely successful attempt to capture the ebb and flow of the battle. The redoubt on Breed's Hill is accurate, although the hill itself is bigger than it should be. It also shows the two main landing points used by the British and places their heavy artillery correctly on Moulton's Hill (top right)

RIGHT: This engraving gives the perspective of the fight from the point of view of someone watching from Noddle's Island. It appears to depict the latter stages of the action, since Charlestown is well alight, but there are too many British warships present: only three, *Lively*, *Falcon*, and *Spitfire*, were in a similar position at this time.

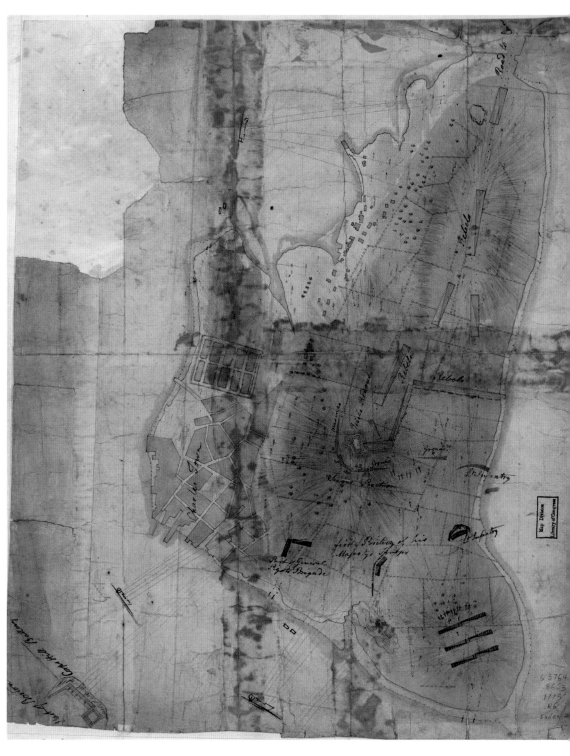

ABOVE LEFT: This map clearly shows that, by building their redoubt on Breed's Hill rather than Bunker's Hill proper, which is shown correctly nearer to the peninsula's neck, the American militia were in real danger of being cut off. In the event the Americans were able to make good their escape across the neck, although they were fired on by two British warships, *Glasgow* and *Symmetry*.

ABOVE: Even though torn and foxed, this is still a more than competent outline of the battlefield's terrain and troop dispositions in the earlier stages of the engagement. It also shows that some American militia units (actually various contingents from Massachusetts under General Israel Putnam) were placed on the real Bunker's Hill and that reinforcements did later cross the neck. These were two Massachusetts regiments and three companies from Connecticut and acted as a rearguard.

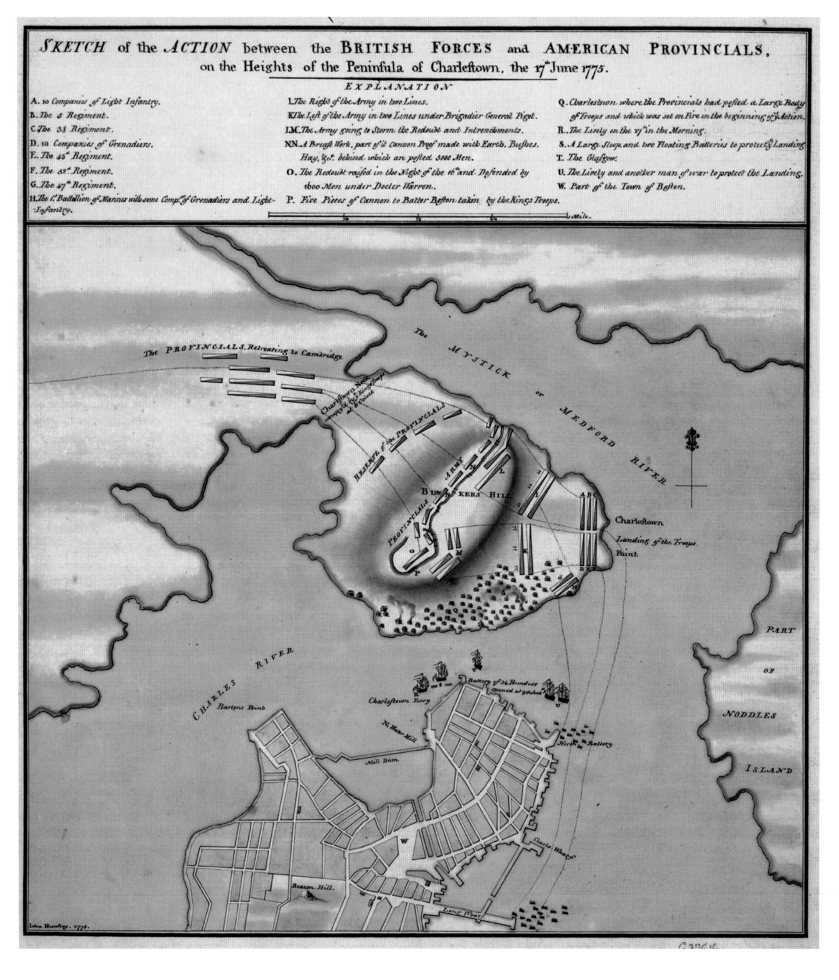

SKETCH of the ACTION between the BRITISH FORCES and AMERICAN PROVINCIALS, on the Heights of the Peninfula of Charleftown, the 17ᵗʰ June 1775.

EXPLANATION

A. 10 Companies of Light Infantry.
B. The 5 Regiment.
C The 38 Regiment.
D. 10 Companies of Grenadiers.
E. The 43ᵈ Regiment.
F. The 52ᵈ Regiment.
G. The 47ᵗʰ Regiment.
H. The 1ᵗ Battalion of Marines with some Comp.ⁱ of Grenadiers and Light-Infantry.

L. The Right of the Army in two Lines.
K. The Left of the Army in two Lines under Brigadier General Pigot.
I.M. The Army going to Storm the Redoubt and Intrenchments.
N.N. A Breaft Work, part of it Cannon Proof made with Earth, Bushes, Hay, &c. behind which are pofted 3000 Men.
O. The Redoubt raifed in the Night of the 16ᵗʰ and Defended by 1600 Men under Docter Warren.
P. Five Pieces of Cannon to Batter Bofton taken by the Kings Troops.

Q. Charleftown where the Provincials had pofted a Large Body of Troops and which was set on Fire in the beginning of ye Action.
R. The Lively in the yᵉ in the Morning.
S. A Large Sloop, and two Floating Batteries to protect ye Landing.
T. The Glasgow.
U. The Lively and another man of war to protect the Landing.
W. Part of the Town of Bofton.

LEFT: This is essentially a colorized version of the map that appears on page 19 and none of the faults evident in the original has been corrected, although a partial street plan of Boston has been introduced and the high ground of Beacon Hill identified.

RIGHT: A panoramic view of Boston and the surrounding countryside, probably sketched from the Dorchester Heights during the prolonged siege that followed the battle. The British recognized that the Charlestown Peninsula was a key point in their defenses and built extensive fortifications along it, including a sizeable position on the real Bunker's Hill.

BELOW: Although clearly a hastily drawn sketch of the battle, this does capture some of the salient detail of the key positions and troops locations. It correctly identifies the locations of the British artillery. The heaviest 12-pounders were positioned on Moulton's Hill (bottom left), while the lighter, more mobile 6-pounders (center) actually moved forward with the assault columns. Some of the latter pieces could not be used, however, as they had been provided with rounds made for their heavier cousins.

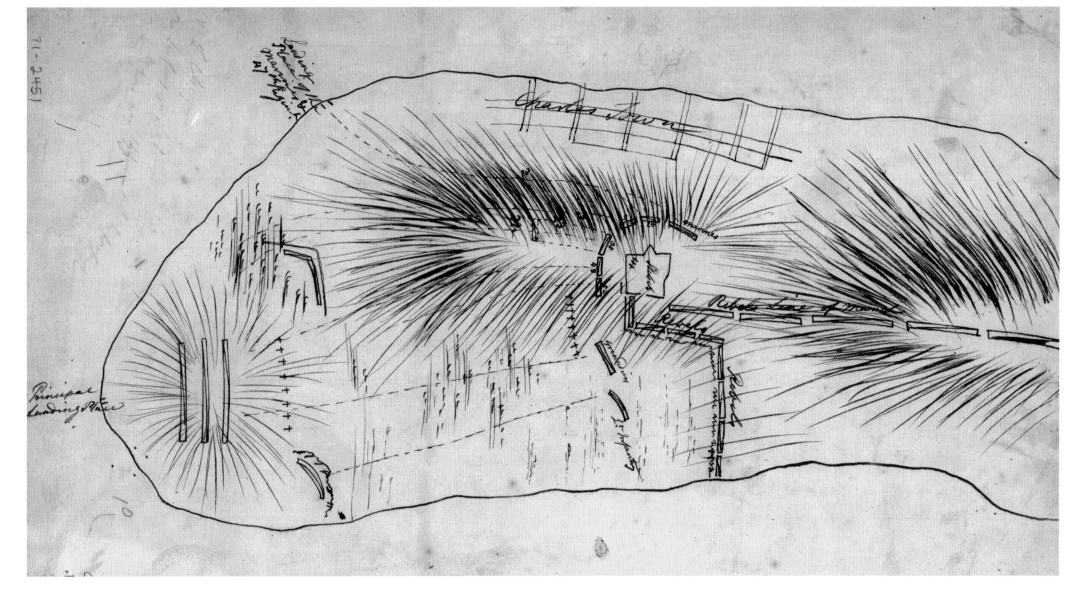

THE BATTLE OF SARATOGA, 1777

DATE: 19 September–7 October 1777

COMMANDERS: (British) Major-General John Burgoyne;
(American) Major-General Horatio Gates

TROOP STRENGTHS: (British) c.7,000; (American) 10,277

CASUALTIES: (British) c.1,200 killed or wounded and 5,895 prisoners;
(American), c.443 killed, wounded or missing

While planning their campaign in the northern colonies in the third year of the American War of Independence, the British opted to drive along the valley of the Hudson River from both the north and south in an attempt to split the rebellious colonies asunder. Major-General William Howe was to move north from New York towards Albany and there link up with Burgoyne, who was to push south from Canada by way of Lake Champlain with an Anglo-German force. While the plan seemed simple enough, it did not take into account the difficulty of the terrain—

largely dense, trackless forest—and, to make matters worse, Howe was not explicitly informed of his part in the mission.

Burgoyne, who had actually devised the operation, began well enough by capturing Fort Ticonderoga on 5 July and mauling the retreating American rear guard at the Battle of Hubbardton two days later. The terrain now began to work against him and he spent the next three weeks slogging his way to Fort Edwards at snail's pace. Burgoyne also finally discovered that Howe was not heading north but had actually gone south to seek out General George Washington. Burgoyne knew that a third column, one that included a large contingent of Native Americans and pro-British Loyalists (Tories to their opponents), under Colonel Barry St. Leger, was also supposedly moving on Albany, so decided to continue pushing down the Hudson Valley.

Burgoyne's situation nevertheless continued to deteriorate as Washington had rushed some 4,600 troops to reinforce those already opposing him. Worse was to follow—a detachment from Burgoyne's command suffered some 900 casualties at the Battle of Bennington on 16 August and a week later St. Leger had to abandon his move on Albany when he was forced to give up the siege of Fort Stanwix. Burgoyne

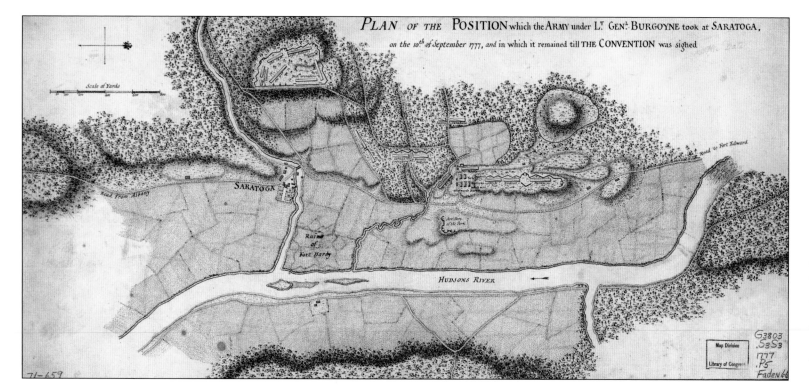

LEFT: The various positions around Saratoga held by the British in the final stages of the campaign, 8–17 October. The Fishkill River, a tributary, runs westward past the town that gave its name to the battle. Burgoyne's multi-national force had taken up various positions to the north of the Fishkill. British and American Loyalists mostly occupied the hill (top left) and the various German contingents defended the ridge (center right). The Battles of Freeman's Farm and Bemis Heights took place some seven miles to the south.

RIGHT: The Saratoga battlefield was heavily forested and dissected by various ravines. This highly detailed map shows the fortified American positions on the Bemis Heights (top left) and further high ground to the west. The British are stretched out along the lower third of the map. Freeman's Farm, though not actually identified here, was located at bottom left.

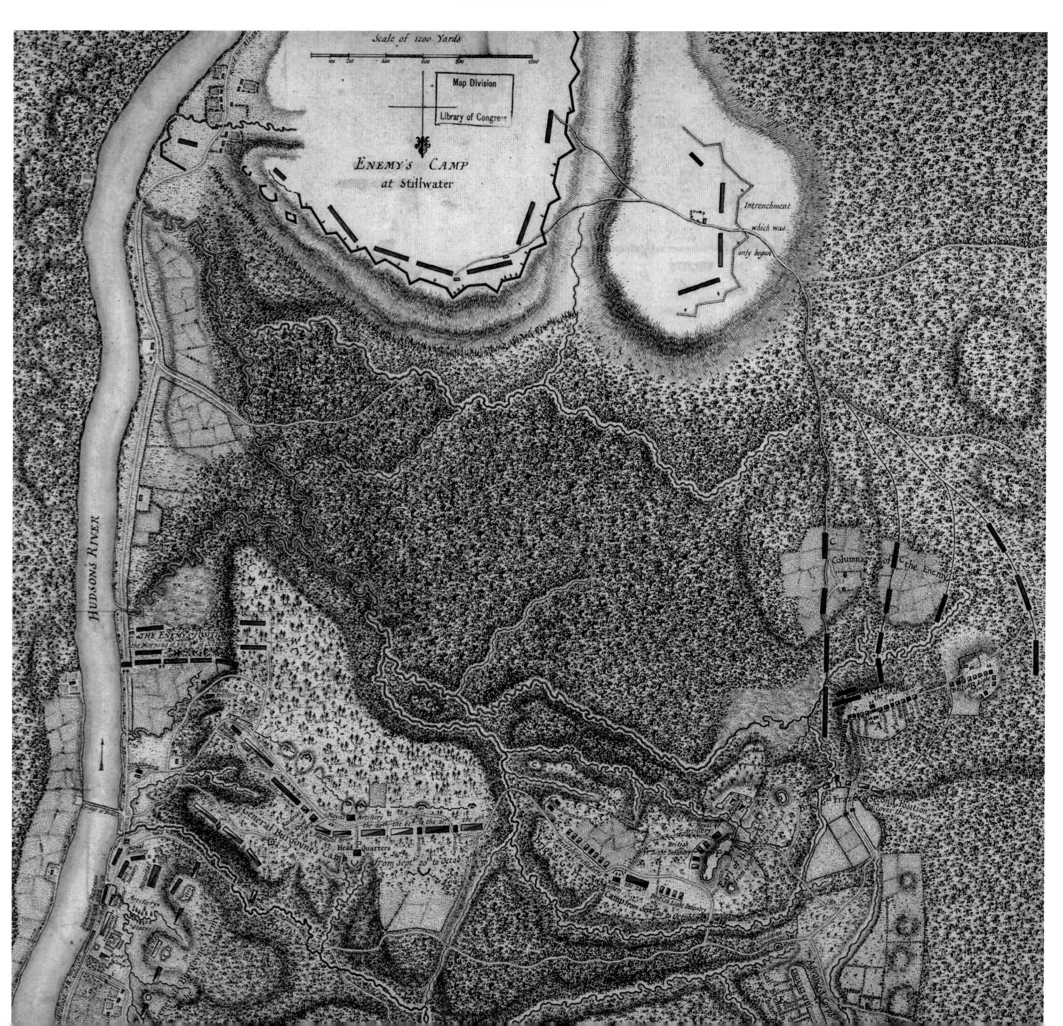

Scale of 1200 Yards

Map Division

Library of Congress

ENEMY'S CAMP
at Stillwater

Intrenchment
which was
only begun

HUDSON'S RIVER

Columns of the Enemy

THE ENEMY'S
the Morning

General Burgoyne's Camp
Head Quarters
from Sept 20th to Octob.

Artillery

British
the Infantry

Camp from Sept 20 to October
British Light Infantry

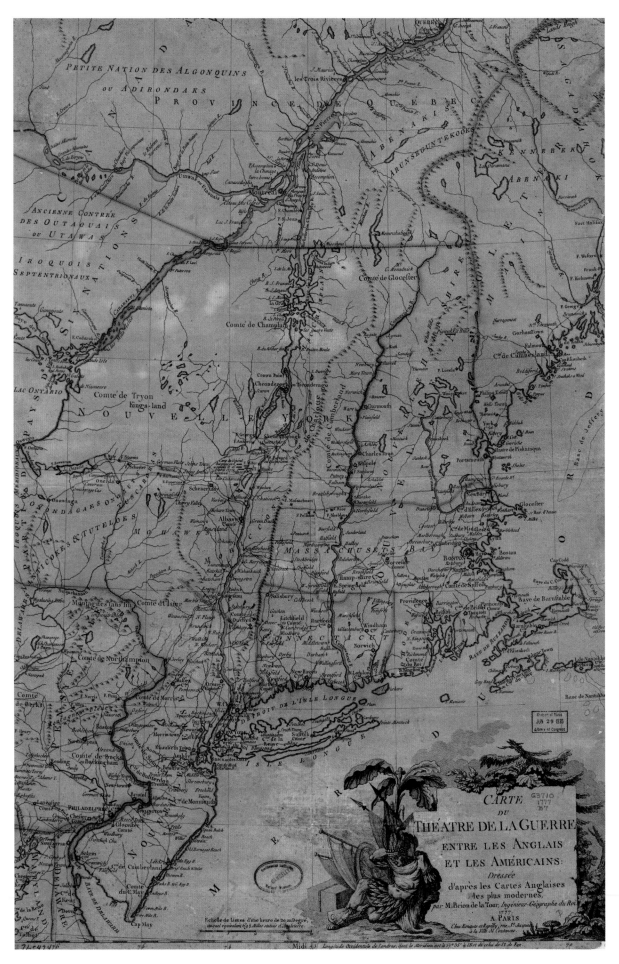

LEFT: The complex British campaign was supposed to use three separate forces to split the rebellious northern colonies in two. One force, Burgoyne's, was to strike south down the Hudson River, making for Albany, while a second under Howe was to push north from New York City to meet Burgoyne at Albany. A third force was to strike eastward from Oswego on Lake Ontario, take Fort Stanwix, and then make for the same rendezvous.

ABOVE: Major-General Horatio Gates, the victor of Saratoga, was actually British-born and had served in the British Army during the French and Indian War (1754–63). He settled in Virginia in 1773 where George Washington later helped secure him the position of adjutant general of the army in 1775.

RIGHT: By the second half of October the British besieged outside Saratoga had fewer than 3,500 effectives under arms and had no hope of salvation. They asked for a truce on the 14th and formally surrendered three days later. Here Gates accepts Burgoyne's sword, which he courteously returned as a mark of respect to the defeated British commander.

was thus isolated and had little option but to press on. He crossed the Hudson River close to Saratoga in upstate New York on 13 September and immediately ran into Gates' troops who were entrenched on Bemis Heights.

The British were determined to attack, so six days later launched an assault on Freeman's Farm on the American left flank. The assault was thrown back with heavy casualties among Burgoyne's command, some 600 in all. Burgoyne now paused to recover and dig his own entrenchments. He had requested Major-General Henry Clinton, who was based in New York with some 7,000 Anglo-German troops, to come to his aid. However, Clinton did nothing more than make a brief sortie in early October that did little to ease Burgoyne's growing sense of isolation and weakness.

The second phase of the battle began on the 7th, when the British made a final desperate bid to turn the American left flank. The Battle of Bemis Heights was a second disaster for the British and they lost around 600 irreplaceable troops. Burgoyne now had no option but to retreat towards Saratoga but his remaining men were soon surrounded by an American force three time his strength. Left with no way out and no hope of rescue, he surrendered on the 17th.

Saratoga was a battle of immeasurable importance since it left the British holding little territory in the northern colonies and, most crucially, it led to France recognizing the independence of the nascent United States.

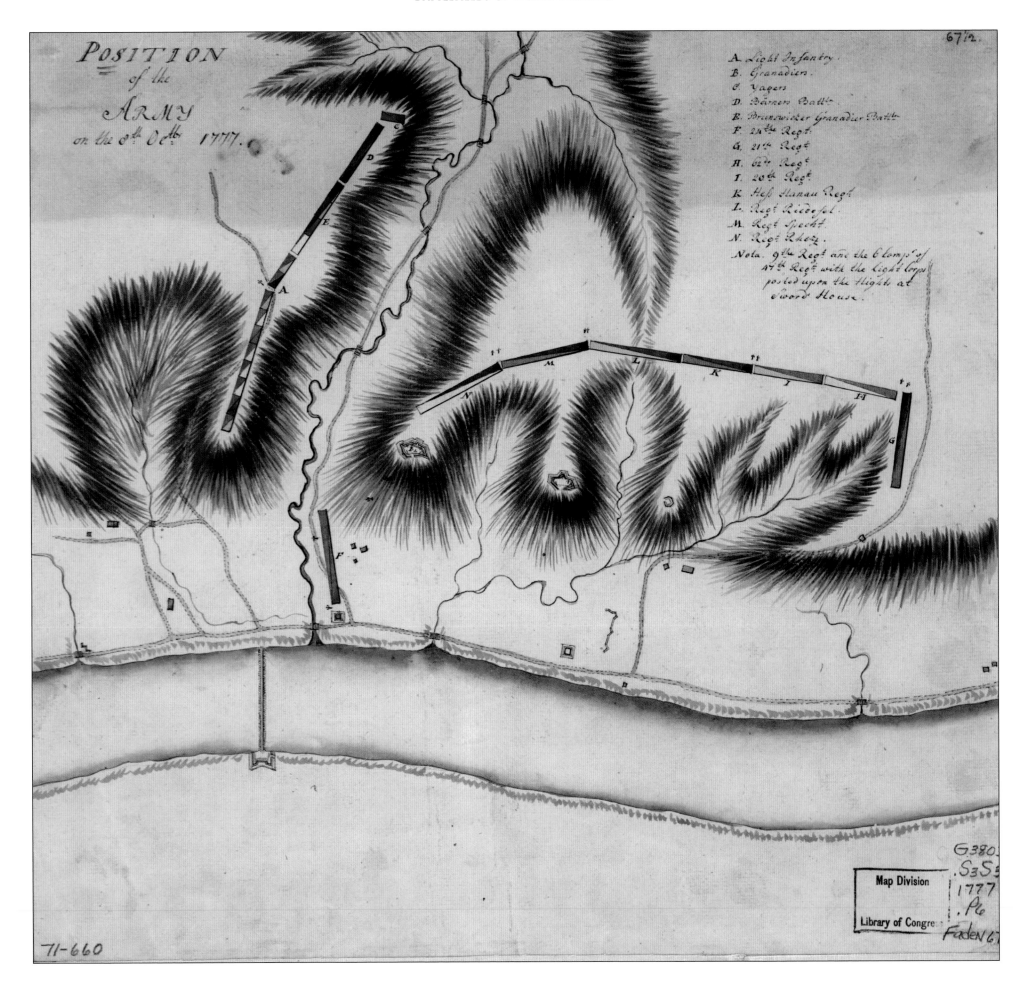

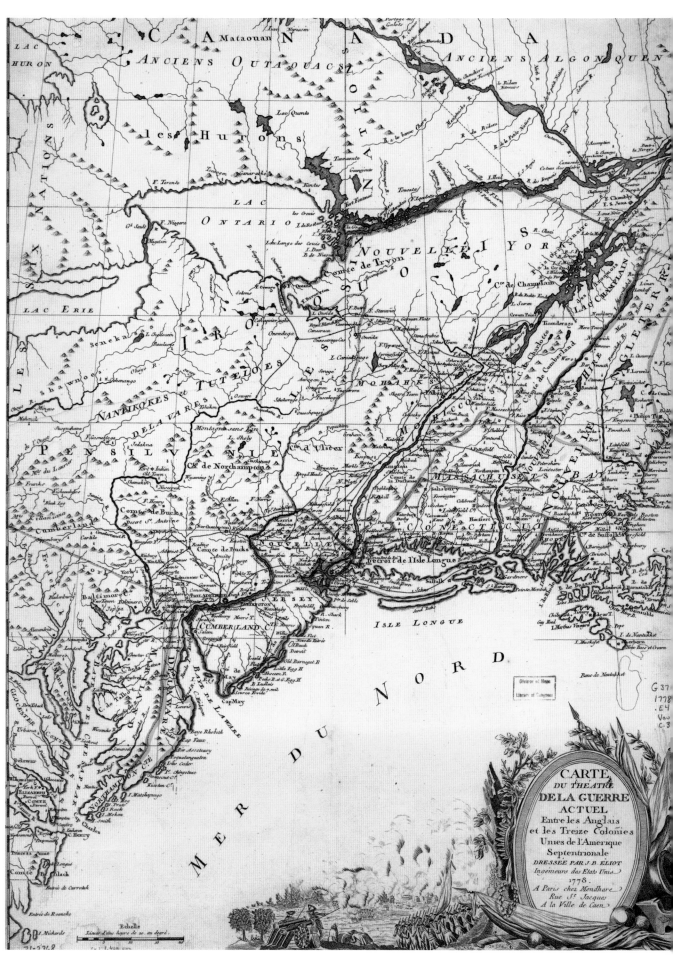

LEFT: Following their defeat at the Battle of Bemis Heights on 7 October the British decided to retreat northward to Saratoga. The movement began at around 0900 hours the next day and, although the American forces did not particularly contest the withdrawal, Burgoyne's men spent a miserable 24 hours as they crossed already sodden ground in the face of blustery winds and sudden downpours. This map shows that, while some units could cross the Fishkill River, others had to briefly remain on its south bank as it soon became too swollen to cross.

ABOVE: British General Sir William Howe was a career soldier of some skill and won several victories during the War of Independence, including Bunker Hill in 1775 and White Plains outside Washington the following year. He was supposed to have supported Burgoyne during the Saratoga campaign of 1777 but actually campaigned in Pennsylvania and he captured Philadelphia on 26 September.

RIGHT: A French map depicting several of the northern colonies that rebelled against the British. New York State (Nouvelle York) is clearly identifiable. The American victory at Saratoga led France to acknowledge the existence of the independent United States.

THE BATTLE OF YORKTOWN, 1781

DATE: 28 September–19 October 1781

COMMANDERS: (American) General George Washington; (French) General Jean Baptiste de Rochambeau; (British) Charles, Earl Cornwallis

TROOP STRENGTHS: (American) 8,845; French 7,800; (British) 8,440

CASUALTIES: (American) 20 killed and 56 wounded; (French) 52 killed and 134 wounded; (British) 156 killed, 362 wounded, 8,087 surrendered, and 44 deserted.

The victory of Franco-American troops at Yorktown was the decisive moment of the American War of Independence and was the end product of France's decision to send a large, well-trained expeditionary force to North America. In the late summer of 1781 Washington, the commander of the Continental Army, had intended to strike against New York City but found its garrison of some 17,000 British troops too tough a nut to crack, and so turned his attention to Virginia. The offensive there was spearheaded by around 2,500 of Washington's troops and backed by a smaller number of French troops led by Rochambeau. The move south in French vessels from Head of Elk near Baltimore and Annapolis began on 20 August and, after overcoming transport problems in the Chesapeake Bay, the force disembarked at Williamsburg northwest of Yorktown during late September.

RIGHT: A wholly fanciful impression of events at Yorktown. The town itself and the immense stone ramparts are shown as far grander than was actually the case. Yorktown was not much more than an overgrown fishing village and the defenses were improvised earth redoubts and other hastily built positions. Equally, no large fleets of warships was directly involved in the battle.

FAR RIGHT: Although comparatively little known, the Second Battle of the Capes of 5–13 September 1781 had a decisive impact on the Yorktown campaign. Victory over a British fleet under Rear Admiral Thomas Graves by French Rear Admiral Francois, Comte de Grasse, insured that the Yorktown garrison was cut off from outside aid. The battle took its name from Capes Charles and Henry, which mark the entrance to Chesapeake Bay, wherein lies Yorktown.

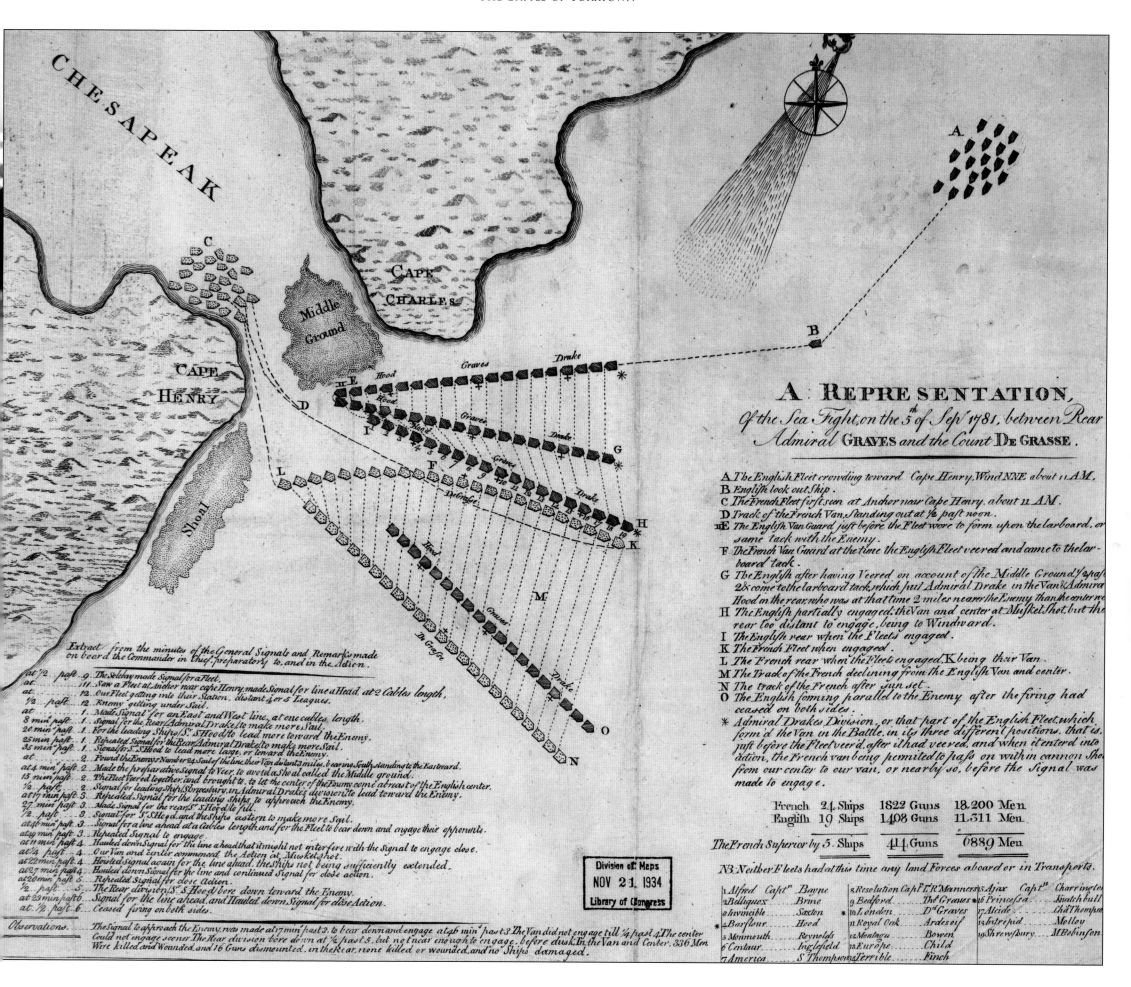

A REPRESENTATION,
Of the Sea Fight, on the 5 of Sep 1781, between Rear Admiral GRAVES and the Count DE GRASSE.

A The English Fleet crowding toward Cape Henry, Wind NNE about 11 AM.
B English look out Ship.
C The French Fleet first seen at Anchor near Cape Henry, about 11 AM.
D Track of the French Van, standing out at ½ past noon.
E The English Van Guard just before the Fleet wore to form upon the larboard, or same tack with the Enemy.
F The French Van Guard at the time the English Fleet veered and came to the larboard tack.
G The English after having Veered on account of the Middle Ground, ¼ past 2 & come to the larboard tack, which put Admiral Drake in the Van & Admiral Hood in the rear, who was at that time 2 miles nearer the Enemy than the center was.
H The English partially engaged: the Van and center at Musket Shot, but the rear too distant to engage, being to Windward.
I The English rear when the Fleets engaged.
K The French Fleet when engaged.
L The French rear when the Fleets engaged, K being their Van.
M The Track of the French declining from the English Van and center.
N The track of the French after Sun set.
O The English forming parallel to the Enemy after the firing had ceased on both sides.
* Admiral Drakes Division, or that part of the English Fleet, which form'd the Van in the Battle, in its three different positions. that is just before the Fleet veer'd, after it had veered, and when it entered into action, the French van being permitted to pass on within cannon Shot from our center to our van, or nearly so, before the Signal was made to engage.

French	24 Ships	1822 Guns	18.200 Men
English	19 Ships	1408 Guns	11.311 Men
The French Superior by	5 Ships	414 Guns	6889 Men

N.B: Neither Fleets had at this time any land Forces aboard or in Transports.

Extract from the minutes of the General Signals and Remarks made on board the Commander in Chief, preparatory to, and in the Action.

at ½ past 9. The Solebay made Signal for a Fleet.
at 11. Saw a Fleet at Anchor near cape Henry, made Signal for line a Head, at 2 Cables length.
at 12. Our Fleet getting into their Station. distant 4 or 5 Leagues.
½ past 12. Enemy getting under Sail.
at 1. Made Signal for an East and West line, at one cables length.
8 min past 1. Signal for the Rear Admiral Drake, to make more Sail.
20 min past 1. For the leading Ships S.S. Hood to lead more toward the Enemy.
25 min past 1. Repeated Signal for the Rear Admiral Drake to make more Sail.
35 min past 1. Signal for S.S. Hood to lead more large, or toward the Enemy.
at 2. Found the Enemys Number 24 Sail of the line, their Van distant 3 miles. bearing South, standing to the Eastward.
at 4 min past 2. Made the preparative signal to Veer, to avoid a Shoal called the Middle ground.
15 min past 2. The Fleet Veered together, and brought to, to let the center of the Enemy come abreast of the English center.
½ past 2. Signal for leading Ship Shrewsbury in Admiral Drakes division to lead toward the Enemy.
at 17 min past 3. Repeated Signal for the leading Ships, to approach the Enemy.
27 min past 3. Made Signal for the rear S.S. Hood to fill.
½ past 3. Signal for S.S. Hood, and the Ships astern to make more Sail.
at 46 min past 3. Signal for a line ahead at a Cables length, and for the Fleet to bear down and engage their opponents.
at 19 min past 4. Repeated Signal to engage.
at 11 min past 4. Hauled down Signal for the line a head that it might not interfere with the Signal to engage close.
at 4 past 4. Our Van and center commenced the Action at Musket shot.
at 22 min past 4. Hoisted signal again for the line ahead. the Ships not being sufficiently extended.
at 27 min past 4. Hauled down Signal for the line and continued Signal for close action.
at 20 min past 5. Repeated Signal for close Action.
½ past 5. The Rear division, S.S. Hood bore down toward the Enemy.
at 25 min past 6. Signal for the line ahead, and Hauled down Signal for close Action.
at ½ past 6. Ceased firing on both sides.

Observations. The Signal to approach the Enemy, was made at 17 min past 3. to bear down and engage at 46 min past 3. The Van did not engage till ¼ past 4. The center could not engage sooner. The Rear division bore down at ½ past 5. but not near enough to engage, before dusk. In the Van and Center, 336 Men were killed and wounded, and 16 Guns dismounted. in the Rear, none killed or wounded, and no Ships damaged.

1 Alfred	Capt. Bayne	8 Resolution Capt. L.R. Manners	15 Ajax	Capt. Charrington
2 Belliquex	Brine	9 Bedford Tho.d Graves	16 Princessa	Kwatchbull
3 Invincible	Saxton	10 London D.d Graves	17 Alcide	Th.e Thompson
4 Barfleur	Hood	11 Royal Oak Ardesoif	18 Intrepid	Molloy
5 Monmouth	Reynolds	12 Montagu Bowen	19 Shrewsbury	M. Robinson
6 Centaur	Inglefield	13 Europe Child		
7 America	S. Thompson	14 Terrible Finch		

The British field commander, Earl Cornwallis, had in late July acted on somewhat unclear orders from Major-General Henry Clinton, the British commander-in-chief, that had effectively led his troops into a dead-end at the tip of the Virginia Peninsula. Matters were made worse when his lines of communication by sea with Clinton in New York were cut after a French fleet under Francois de Grasse defeated a rival British force under Admiral Thomas Graves at the Battle of the Capes, fought between 5–9 September. Cornwallis' isolation was thus completed when both Washington and Rochambeau, along with a power train of siege artillery arrived outside Yorktown.

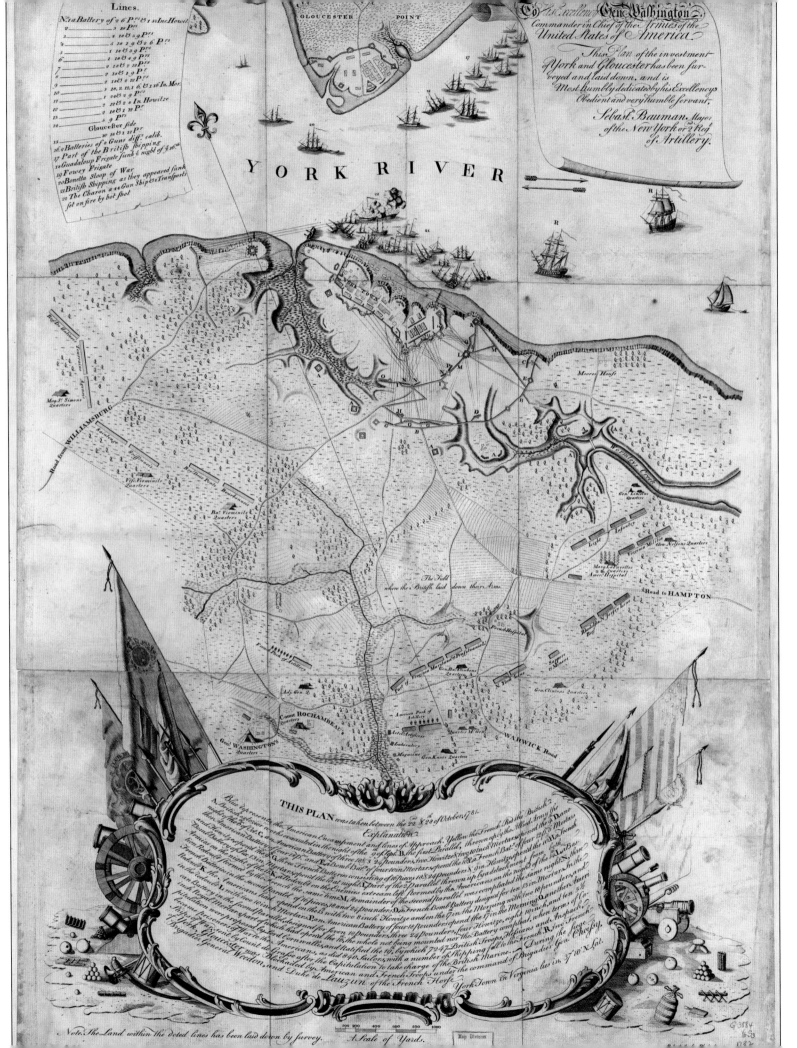

LEFT: A mostly accurate map of the battle area, although the excessive number of British warships present owes more to the maker's imagination than reality since there were only two present, *Charon* and *Guadalope*, the former being sunk by French artillery fire on 10 October. Note also that the British had a garrison across the York River on Gloucester Point (top). This was kept largely penned in by a mixed force of US and French troops after a brief skirmish on 3 October.

LEFT: The terrain around Yorktown ensured that the US-French siege operations could be prosecuted only from the southeast since the ground to the town's west was dissected by various creeks. However, several bastions and redoubts mounting artillery were constructed to block any potential escape route for the British garrison.

ABOVE: Washington was made general and ordered to command the forces besieging Boston in June 1776, and thereafter commanded the continental Army throughout the war.

RIGHT: Washington's forces are feted by ecstatic crowds. Yorktown did not end the war immediately but it insured that the British could never be the victors. Congress finally proclaimed an end to the war on 17 April 1783.

Cornwallis withdrew to Yorktown's inner ring of fortifications on 30 September, an error that allowed the Franco-American artillery to be pushed forward so that the whole of the town's defenses could be brought under direct fire. The preparatory bombardment opened on 9 October and five days later two key British redoubts, Nos. 9 and 10, were stormed in the mid-evening by American and French detachments at the point of the bayonet. A British counterattack was thwarted two days later. Cornwallis, who was well aware of his perilous situation, contemplated an evacuation across the York River to Gloucester Point but the scheme was abandoned due to the sudden arrival of a storm on the 16th.

The end was clearly in sight and the following day Cornwallis began negotiations regarding the terms of his capitulation. Washington, for his part, allowed two days for written proposals to be discussed but steadfastly demanded total surrender. The British troops marched out of Yorktown on the 19th, paraded past their besiegers, and laid down their arms before marching into captivity. The war would continue until a peace treaty was agreed in late November 1782 and the last British troops did not leave New York for another twelve months, but United States independence was assured at Yorktown.

THE BATTLE OF NEW ORLEANS, 1815

The somewhat misnamed Anglo-American War of 1812 had actually ended when this battle took place, since the peace terms, the Treaty of Ghent, had been signed on 24 December 1814, but it took until the second week of February 1815 for the news to reach North America. Thus, still believing that the war was continuing, the British mounted their operation against New Orleans from Jamaica and their troops, mostly veterans of the Napoleonic Wars, embarked on transport vessels on 26 November. The plan was to take the city-port and thereby take control of the valley of the Mississippi. However, rumors of the invasion were rife in the United States and Jackson himself traveled to Louisiana on the 22nd to investigate the state of the city's defenses.

RIGHT: The British forces at the Battle of New Orleans were sandwiched between the Mississippi River on their left flank and a dense cypress swamp on their right, and were therefore forced to attack on a comparatively narrow front and were thus unable to avoid a costly frontal assault on Line Jackson.

FAR RIGHT: A dramatic if somewhat inaccurate rendition of the fighting depicts General Andrew Jackson on the parapet of the improvised defense line, helping to turn back an attack by the British 93rd Regiment of Foot, the Sutherland Highlanders. In reality, the Highlanders, who wore tartan trews and bonnets rather than kilts and bearskins during the battle, never actually attacked Line Jackson due to a misunderstanding over orders that left them standing to attention under a murderous fire.

The British dropped anchor on 13 December, landing in the Lake Borgne area to the east of New Orleans, and pushed forward to within seven miles of the city. Jackson reacted immediately by imposing martial law and ordering the building a defensive position, Line Jackson, along the disused Rodriquez Canal. Here his right flank rested on the Mississippi and his left was anchored on a cypress swamp. This was manned by a motley mixture of a few regular troops, state militia, and various volunteers from elsewhere.

Jackson ordered a raid on his opponents under cover of darkness on 23–24 December but it was easily repulsed, while a British probe towards his defenses four days later was repelled by concentrated artillery fire. Pakenham opened his own barrage on 1 January but this had little impact on Line Jackson, and the British commander had to delay his attack until all of his troops were concentrated. He decided to divide his force into three groups, with those on the left and right making the main assault against Line Jackson, and the center held as a reserve. A subsidiary

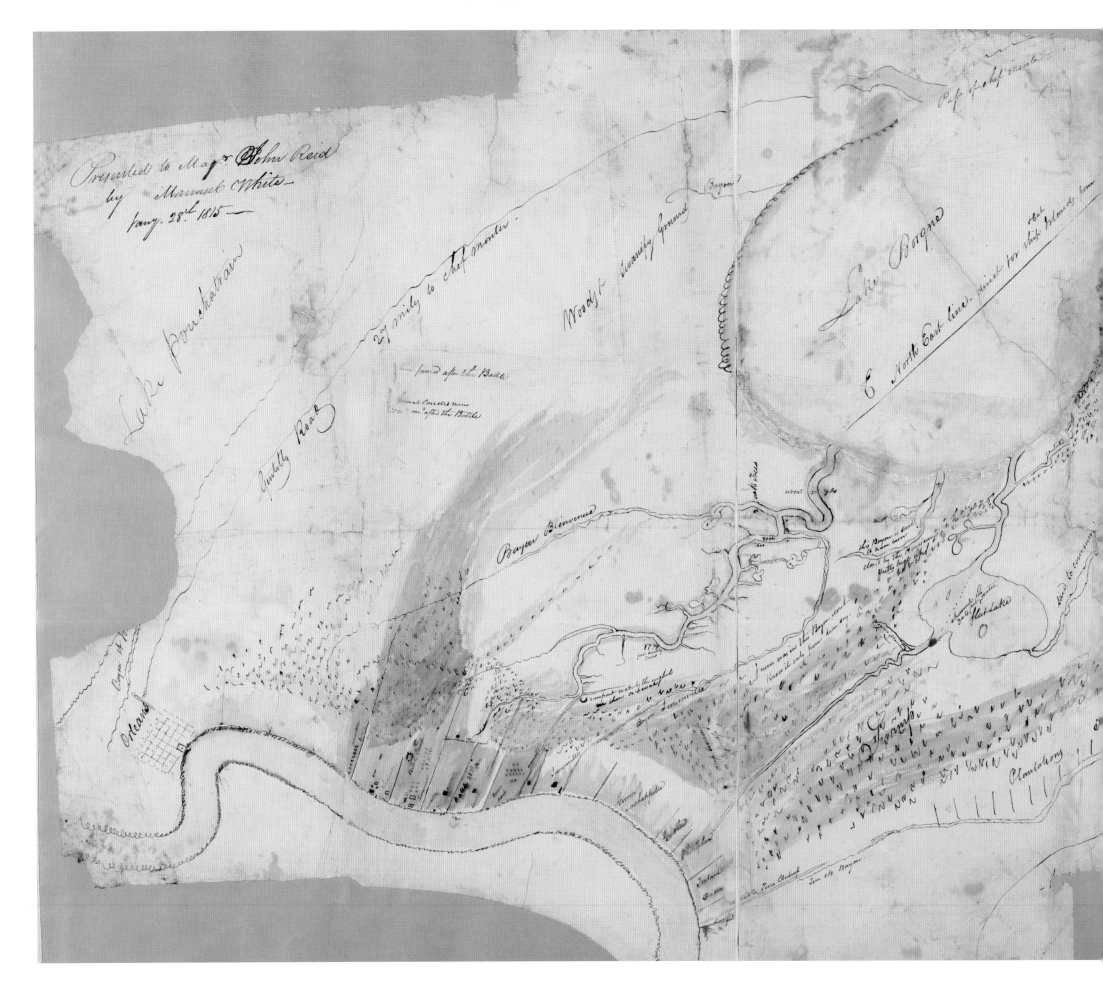

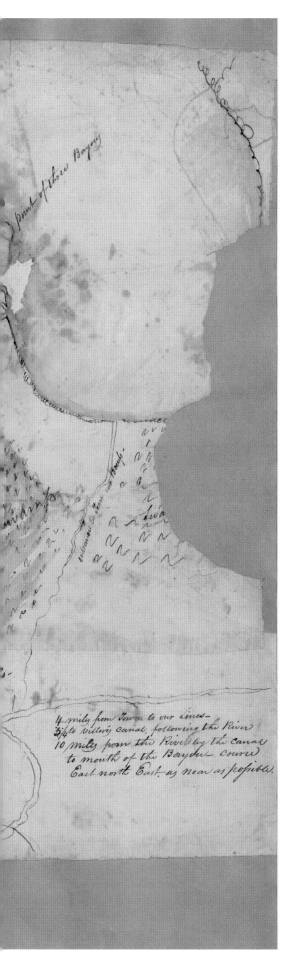

ABOVE: A somewhat romanticized and inaccurate depiction of the death of Major-General Thomas Pakenham in the latter stages of the fighting at New Orleans. He was hit three times in rapid succession by grapeshot and bullets and actually died in the arms of his aide, Major Duncan Macdougal, after falling from the latter's horse. Pakenham's body was later eviscerated and packed in rum for the journey back to its final resting place in St. Paul's Cathedral, London.

LEFT: A map showing the area of operations around New Orleans. The British landed at Fisherman's Village on the shores of Lake Borgne (right) on 22 December 1815, and then swung south and east to move along the east bank of the Mississippi River. Jackson reacted to the landings by building a series of defensive positions across the British line of march.

attack was also made on the opposite bank of the Mississippi. When the troops advanced against the line, they were hit by a hurricane of rifle and cannon fire that cut them down in droves. Those on the right failed to storm the American ramparts but a few men on the left did briefly break into one bastion on the banks of the Mississippi. Yet the moment was soon lost and all of them were either killed or captured.

Pakenham went forward to attend to matters on the right but was hit by a piece of grapeshot, then received a bullet in one arm, and was finally felled by a second. He died in the arms of an aide. Thereafter the British began to withdraw from the battlefield, even though the subsidiary assault had been successful. Jackson did not pursue, but maintained his position behind the Rodriguez Canal.

New Orleans was the first unequivocal major US victory of the conflict and it made Jackson a popular hero, one who would be propelled into the president's office in 1829.

THE SIEGE OF THE ALAMO, 1836

DATE: 23 February–6 March

COMMANDERS: (American) James Clinton Neill, then Lieutenant-Colonel William Barret Travis; (Mexican) General Antonio Lopez Santa Anna Perez de Lebron

TROOP STRENGTHS: (American) 189 but possibly more; (Mexican) 1,700 infantry and 239 cavalry (final assault)

CASUALTIES: (American) 189 killed; (Mexican) c.500–600 killed and unknown number wounded

American settlers in Texas rose against the ruling Mexican authorities on 30 June 1835 demanding the creation of an independent republic. There were various clashes between the two over the next several months but it was not until the following year that the Mexican president and self-styled "Napoleon of the West," Santa Anna, was ready to lead his army into Texas to put down the revolt. There were two viable routes into the fledgling republic—the southerly Atascosito Road and the more northerly El Camino Road. Santa Ana sent a small force along the southerly route while he himself accompanied the main invasion army along the northerly route, which began at the Presidio del Rio Grande, a frontier fort on the river of the same name.

The overall commander of the still-forming Texan army, General Sam Houston,

LEFT: General Martin Perfecto de Cos was the brother-in-law of the Mexican president, General Antonio de Santa Anna. In 1835 he led some 1,400 troops into Texas to re-impose Mexican authority. The first clash came on 2 October when a small detachment of Cos's troops under Lieutenant Francisco Castaneda, trying to repossess an old cannon, was fired on by pro-independence Texans a little to the south of Gonzales.

RIGHT: In 1727 the Spanish colony to the north of Mexico became the province of Tejas or Texas, a name derived from a local tribe of Native Americans, but little settlement took place. Mexico gained its independence from Spain in 1821, and by 1835 around 28,000 Americans had settled in Texas. Texan calls for autonomy were rejected by Mexican President Santa Anna, who sent in troops to control the increasingly unruly pro-independence movement.

LEFT: Following the siege of the Alamo, the Battle of San Jacinto took place on 21 April 1836, when a Texan force under General Sam Houston inflicted a crushing defeat on the Mexicans. Here, Houston (left) accepts the surrender of Generals Santa Anna and Cos. The speech bubbles—Houston declaring that the pair are "bloody villains" and Santa Anna stating "Me no Alamo!!"—effectively reflect the wave of anti-Mexican feeling that swept the United States after the massacre at the Alamo mission station on 6 March and the less well-known but even more unjustified murder of 342 Texan prisoners, including their commander, Colonel James Walker Fannin Jr., at Goliad some three weeks later.

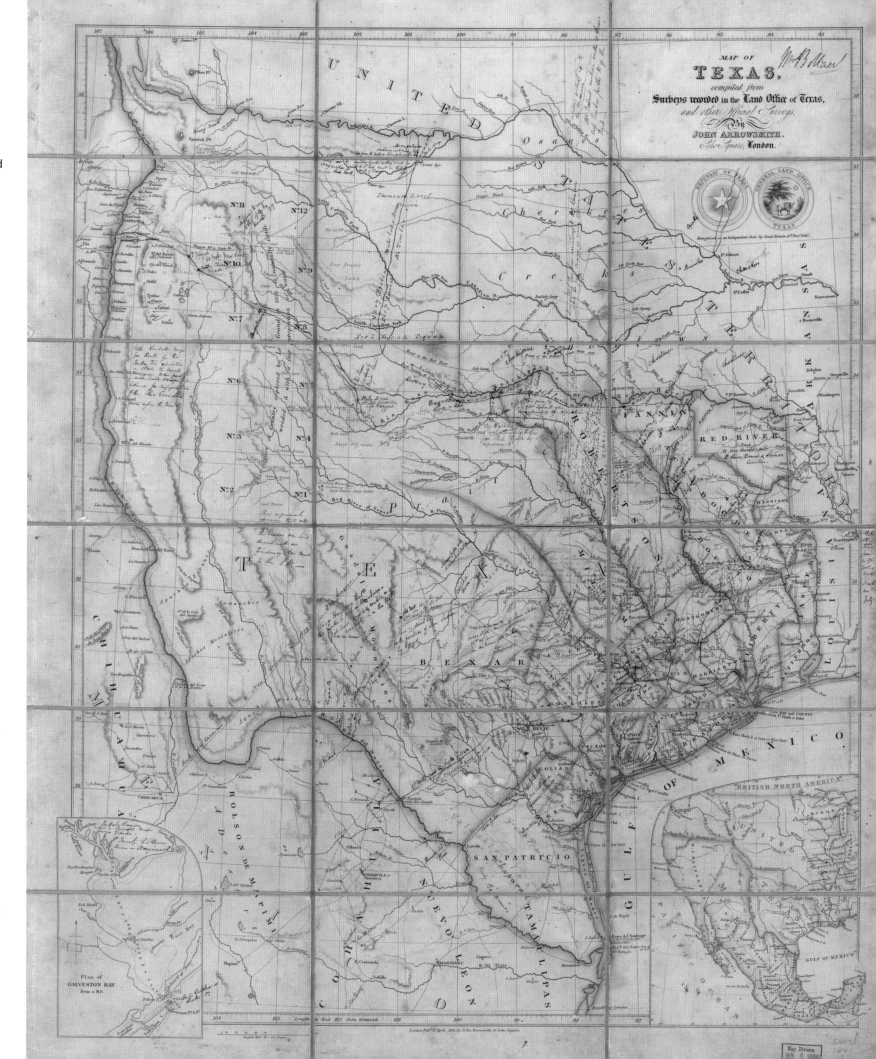

RIGHT: General Antonio Lopez Santa Anna Perez de Lebron was born into the Creole middle class and served in the Spanish Army in 1810, but became attracted to the Mexican independence movement. Independence was achieved in 1821 and Santa Anna then led a coup against the movement's increasingly despotic head, General Agustin de Ituribe—the self-styled Agustin I—the following year. He then became governor of Veracruz province and in 1829 rebuffed an attempt by Spain to re-establish control over its former colony at Tampico. The so-called "Hero of Tampico" became president in 1833.

had few troops as yet available and more than anything else he needed time to gather and train them. He was also well aware that there were only two routes into Texas and decided to place holding forces on both. Troops were sent to the Presidio del Bahia at Goliad in the south and to the dilapidated Alamo mission at San Antonio de Bexar in the north. The latter were a mixture of volunteers, former professional soldiers, and backwoodsmen, including the renowned Jim Bowie and David Crockett. Santa Anna crossed the Rio Grande on 16 February and reached the Alamo, whose makeshift defenses were barely completed, a week later.

LEFT: Sam Houston served in the US Army before resigning to practice law. He became a politician, serving in the House of Representatives from 1823. He was a heavy drinker, but built a hero's reputation, during the Texan War of Independence.

ABOVE: David Crockett was a living legend in 1836, renowned as a fighter of Native Americans, as a bear hunter, and a three-term congressman. Contrary to such popular images, Crockett preferred to wear conventional clothes rather than buckskins.

RIGHT: A glamorized depiction of the final moments of the siege, showing Mexican troops breaking into the mission's church. It somewhat misrepresents the reality of the event. Jim Bowie (center left) did not die in the church but was actually killed in his bed in a room adjacent to the Alamo's main gate. He had been suffering from a malaise variously referred to as "hasty consumption" or "typhoid pneumonia." David Crockett and six or seven of his men did briefly survive the storming of the church to be taken prisoner, but were swiftly executed on Santa Anna's orders.

The original Mexican plan was to either bombard the Alamo's insubstantial adobe walls into powder or starve the garrison into surrender. Thus the Texans there came under bombardment from the outset but, because the Mexican ordnance consisted of light cannon, the damage was not as great as if siege artillery had been available. Travis, who had taken over after Neill had left to attend to family business on the 14th, expected substantial help from Houston but the only reinforcements that arrived were the thirty-two mounted men who cut their way through the Mexican cordon on 1 March.

Matters deteriorated fast for the Texans. Three days later the Mexicans established a battery close to the Alamo's north wall, which was soon close to collapse. The same day Santa Anna held a council of war and declared that the mission would be taken by storm on the 6th, much to the surprise of his commanders who felt a costly assault was unnecessary. Nevertheless, the various assault detachments stood to at 0300 hours and the attack commenced at 0530, with troops advancing against the fort's northern,

northwestern, and eastern sides, as well as against a low parapet close to its church.

The barely awake defenders roused themselves, and their rifle and cannon fire inflicted terrible losses on the attackers. The assault in the east and against the parapet got nowhere initially but those thrown against the weakened northern wall turned the tide. Mexican troops eventually scaled it and entered the courtyard in strength. Some of the garrison, around seventy-five men, now tried to escape but were cut down by the waiting Mexican cavalry. Those still fighting retreated to the Alamo's long barracks and church but were methodically killed. Six or seven prisoners were taken but Santa Anna had them immediately executed.

Santa Ann's victory at the Alamo was short-lived. His forces were roundly beaten at the Battle of San Jacinto on 21 April by the troops that Houston had been able to raise in the weeks since he crossed the Rio Grande. As Americans advanced to attack, many shouted "Remember the Alamo. Remember La Bahia." Texas gained its independence and was recognized as an independent republic by the United States on 4 July.

ABOVE: Generally accepted estimates of the number of defenders who died during and immediately after the fierce fighting at the Alamo vary from 182 to 189, although some researchers have suggested that over 250 might be more accurate.

THE BATTLE OF PALO ALTO, 1846

DATE: 8 May 1846

COMMANDERS: (US) General Zachary Taylor;
(Mexican) General Mariano Arista

TROOP STRENGTHS: (US) 2,228; (Mexican) 3,500

CASUALTIES: (US) 9 killed, 44 wounded, and 2 missing; (Mexican) 102 killed, 129 wounded, and 26 missing

The US-Mexican War erupted over a border dispute in newly independent Texas. Mexico claimed that it should run along the Rio Grande, and the Texans that it should follow a line north of the Nueces River. The friction between the two was exacerbated by a growing expansionist movement in the United States that became manifest after the election of pro-expansionist President James Polk in 1844. US forces struck first and on 24 March General Taylor pushed southwards with some 3,500 troops to establish an earthwork base, Camp Texas, close to the Rio Grande in southwest Texas, opposite the Mexican town of Matamoros.

The first clash came on 25 April, when Mexican cavalry crossed the river near

LEFT: A dramatic interpretation of the fighting at Palo Alto shows General Zachary Taylor (right) mounted on his favorite horse, "Old Whitey," urging his troops to attack through a somewhat bizarre landscape. The battlefield was, in fact, largely open plain with peripheral patches of thick chaparral, while the fighting was really decided by US cannon fire and not at the point of the bayonet.

RIGHT: Various sketch maps of the fighting in southwest Texas 8–9 May 1846. These are (counter-clockwise from top left) Taylor's dispositions on the eve of the Battle of Palo Alto on the 8th, the battle itself, the brief Battle of Resaca de la Palma fought the following day, and the defensive positions adopted by the US forces during the intervening night.

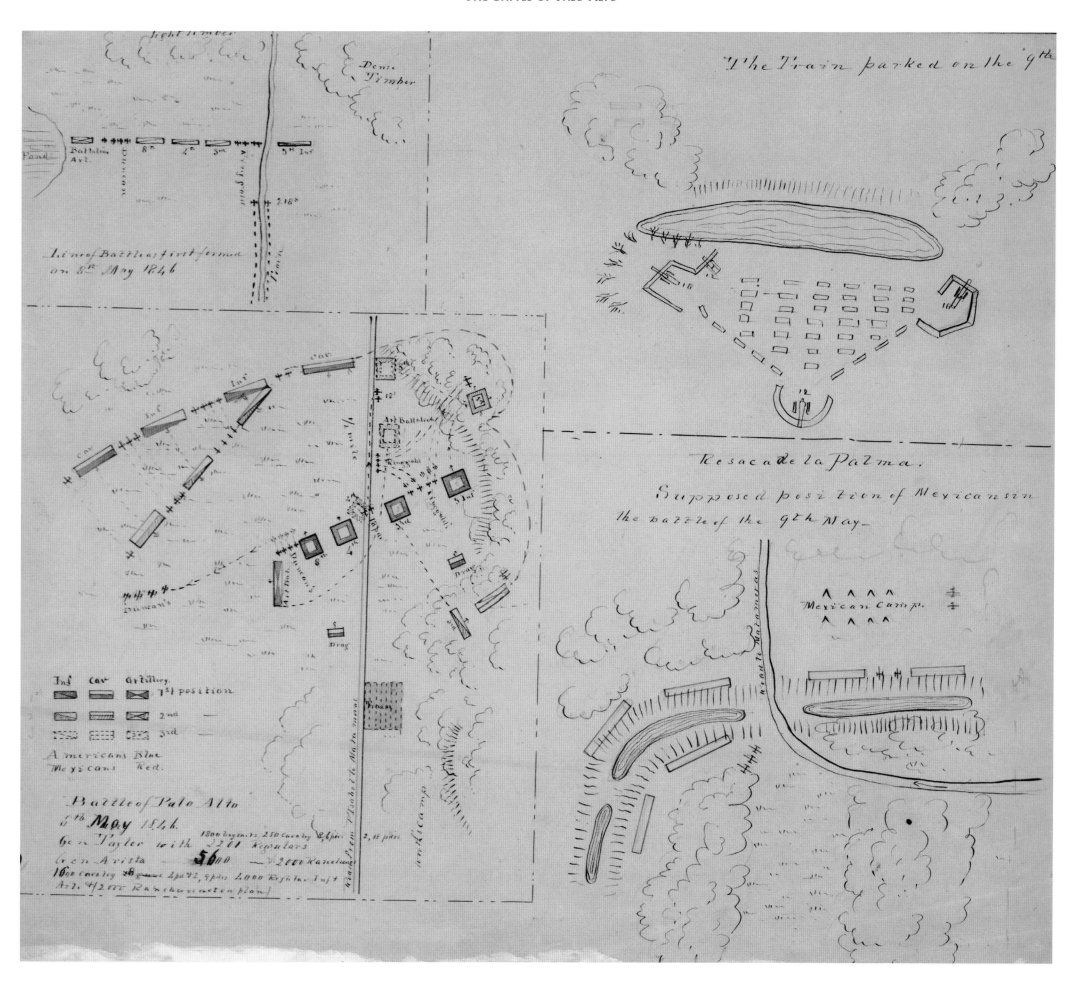

ABOVE: A view of the fighting, looking from the US line toward that of the Mexican forces under General Mariano Arista. The artist has rightly captured the decisive importance of artillery fire during the battle. Taylor has three of the US Army's five batteries of fast-moving "flying artillery" at Palo Alto and these proved decisive as they moved between areas of danger throughout the engagement, with devastating impact. Among those US officers present at Palo Alto destined for greater things was 23-year-old Lieutenant Sam Grant, better known as Ulysses S. Grant, of the US 4th Infantry Regiment.

RIGHT: A statement of US intent before the outbreak of war—neat rows of tents on the shore of the Gulf of Mexico, to the north of Corpus Christi, Texas, mark Taylor's winter headquarters. This base of operations was established in October 1845, some five months after the United States had annexed Texas, and it was not until the following spring that the general led his 3,000 or so men against Mexico with orders from President James Polk to "defend the Rio Grande."

Matamoros and overwhelmed a small US cavalry detachment. Taylor immediately declared that the war had begun, although a formal US declaration was not announced until 13 May. He also demanded volunteers from Texas and Louisiana and on 1 May moved the bulk of his command to Point Isabel on the Gulf Coast, some twenty-five miles east of Matamoros, to protect his main supply base. Camp Texas was left in the hands of Major Jacob Brown and 500 troops.

On 1 May Arista led some 6,000 troops across the Rio Grande and pushed on Camp Texas, which was besieged between the 3rd and 8th. The camp repelled the attacks but Brown was killed—the site became Brownsville in his honor. Taylor, having fortified Point Isabel, now rushed back to Camp Texas. The two forces ran into each other at Palo Alto (Tall Timber), some five miles north of the camp, on the 8th and the battle took place on a piece of mostly level and open ground.

The Mexican line stretched for around a mile; the infantry units were interspersed with heavy cannon and Arista's cavalry were posted on the flanks. Despite their clear numerical advantage, the Mexicans did not attack during the morning. Battle commenced in the early afternoon, when Arista's artillery opened a fierce if largely ineffective barrage. Taylor took the initiative and ordered his troops and artillery to advance. They halted just beyond musket range and the artillery opened up, concentrating on the massed ranks of Mexican infantry. The cannon's grapeshot did terrible damage but Arista ordered his men to stand fast for an hour. Then his cavalry

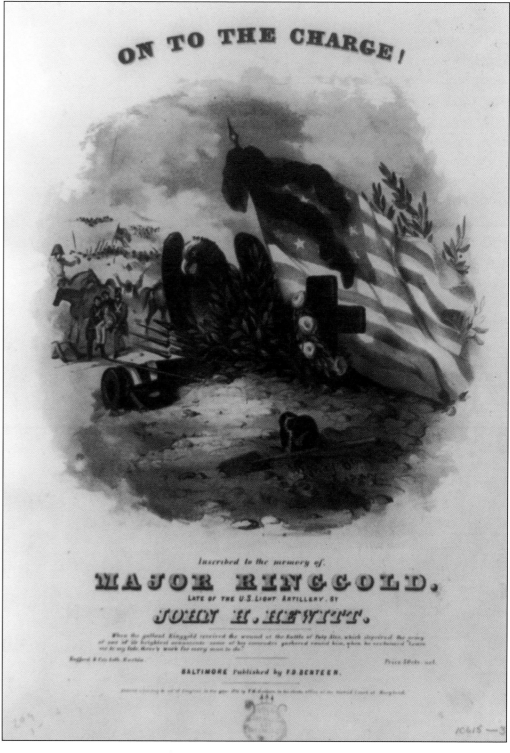

LEFT: Virginia-born and Kentucky-raised Taylor captured in profile. "Old Rough and Ready," as he was nicknamed because of his scruffy dress and lack of pretensions, had considerable military experience before the US-Mexican War. He fought in various campaigns against a number of Native American tribes, most notably crushing the Seminoles of the Florida Everglades at the Battle of Lake Okeechobee in 1837.

ABOVE: A handbill commemorating Major Sam Ringgold, the artillery officer whose development of "flying artillery" played such an important role in the US victory at Palo Alto. These batteries consisted of small bronze cannon slung low between oversize wheels for rapidity of movement, but they had never been used in anger before the battle. Ringgold was mortally wounded after being hit in both thighs by Mexican cannon fire.

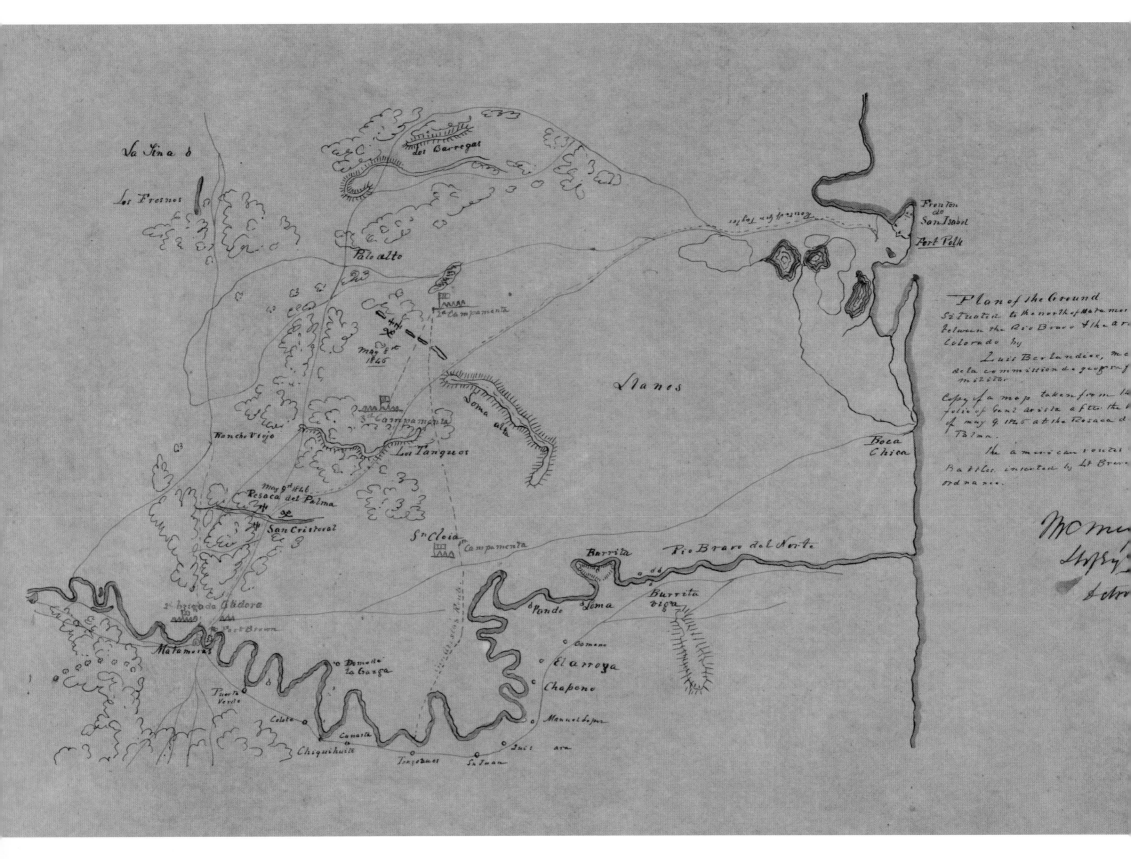

attacked but were repulsed by some of Taylor's infantry, who were ably aided by his highly mobile "flying artillery."

The battle ebbed and flowed throughout what remained of daylight. It was increasingly dominated by an artillery duel, which the better-trained US artillery crews won. The action ended as night fell. Taylor was left holding the original Mexican position, while Arista had retreated to a supposedly impregnable position at Resaca de la Palma, but one where his army would be utterly defeated on the 9th. Taylor went on to invade Mexico proper in August, taking Monterrey in September and winning the Battle of Buena Vista the following 22–23 February, a victory that ended the campaign in northern Mexico.

LEFT: The nothern campaign of the war saw Taylor win three great battles at Palo Alto, Resaca de la Palma, and Monterrey in 1846, and a fourth at Buena Vista in February 1847. Total US casualties were 267 killed, 456 wounded, and 23 missing, while Mexican losses were around 500 dead and 1,000 wounded.

RIGHT: US forces later launched a major invasion of central Mexico with the intention of capturing Mexico City. The invasion troops came ashore near Vera Cruz on 9 March 1847 and invested the port. This map shows them taking Vera Cruz by storm (center top), whereas it was actually bombarded into submission between the 22nd and 28th. Mexico City finally fell on 14 September after several hard fights.

BELOW: A splendidly (but inaccurately) accoutered Taylor depicted directing operations.

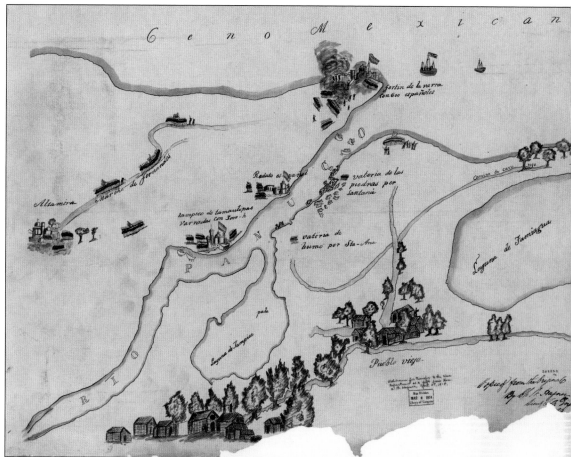

ABOVE: An illustration purporting to capture the action at Palo Alto, but one showing troops dressed in largely imaginative uniforms. The cavalryman has a passing resemblance to a US dragoon—the 2nd Regiment was present at the battle—but the Mexican artillery officer he is aiming his pistol at would have worn a dark blue jacket and trousers.

THE FIRST BATTLE OF FORT SUMTER, 1861

DATE: 12-14 April 1861

COMMANDERS: (Union) Major Robert Anderson; (Confederate) Brigadier-General Pierre Gustave Toutant Beauregard

TROOP STRENGTHS: (Union) 48 guns plus 127 men; (Confederate) 30 cannon and 18 mortars plus c.500 men

CASUALTIES: (Union) 2 killed and 4 wounded; (Confederate) 4 wounded

The election on 6 November 1860 of President Abraham Lincoln, a Republican with anti-slavery sentiments, led seven southern states to secede from the Union to form the Confederate States of America. South Carolina led the way on 20 December, and the Confederacy's first and only president, Jefferson Davis, was elected on 9 February 1861. The outgoing US president, James Buchanan, made little effort to prevent the seizure of US military installations across the seceding states, but a pair of regular officers acting on their own initiative saved two especially important strong-points. One was Fort Pickens in Florida's Pensacola Bay, while the other was Fort Sumter, a somewhat incomplete pentagonal brick edifice some three miles offshore in

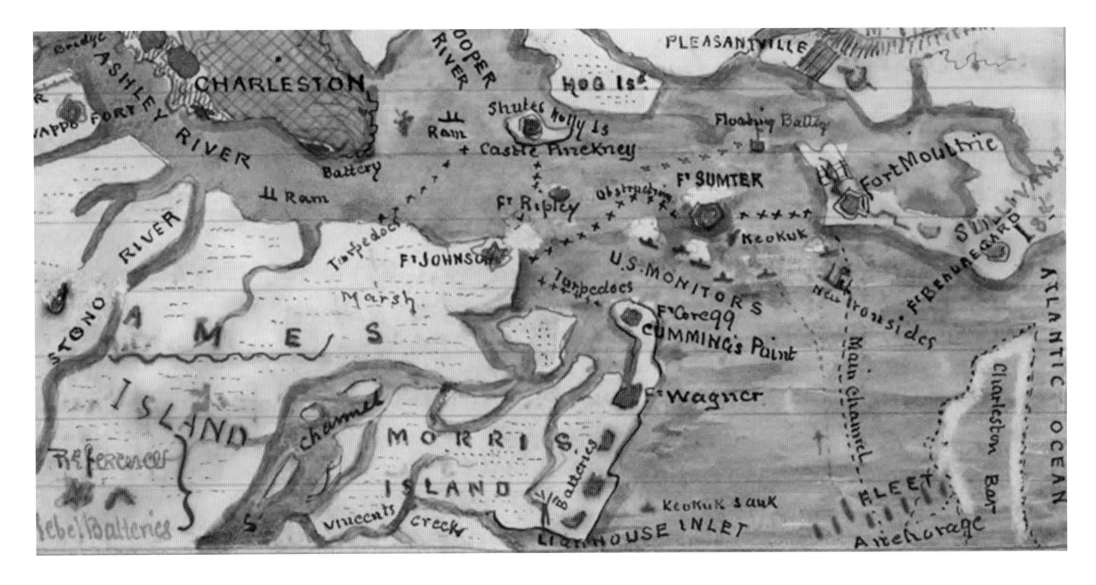

RIGHT: An architectural plan of Fort Sumter. It was erected on a man-made island, one constructed from some 70,000 tons of granite and other rock, which was placed on a shoal at the entrance to Charleston Harbor. Work began in 1829 and was mostly if not entirely completed by 1860.

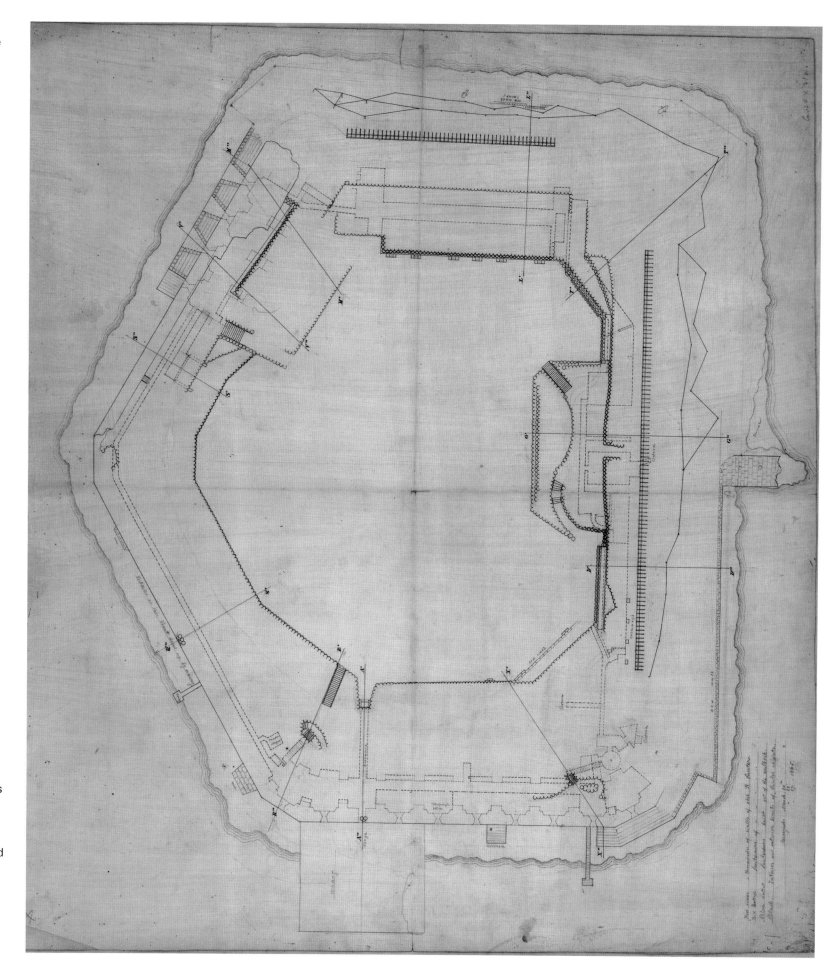

LEFT: Union forces made several attempts to recapture Fort Sumter and subjugate Charleston after the former had fallen into Confederate hands. The first major attack, which is depicted in this map, came on 7 April 1863 and consisted of an attack by nine ironclad warships under Flag Officer Samuel DuPont. When his flotilla sailed towards their target, they were hit by sustained and ferocious artillery fire from Forts Sumter and Moultrie. Five of DuPont's vessels suffered significant damage and one, USS *Keokuk*, was so badly crippled that it sank the next day. This fiasco led to a change in strategy. Henceforth, both Union land and naval forces were committed to capturing Charleston.

the harbor of Charleston, South Carolina. The officer responsible for the latter coup was Kentucky-born Robert Anderson.

In January 1861 Buchanan had sent a boat, the *Star of the West*, loaded with food and ammunition to Sumter but it had been forced to withdraw when fired on from the South Carolina mainland. Thereafter the state's authorities announced that no supplies of any kind were to be allowed to reach Anderson. Lincoln officially took office on 4 March and vowed in his inaugural address to "hold, occupy and possess" federal property in the Confederacy. Anderson was by this stage acutely short of supplies so the newly incumbent president, who had no intention of abandoning the fort but equally did not want to initiate hostilities by taking military action, ordered that a supply boat be dispatched in early April.

The Confederacy's response came on the 7th after they had been formally informed of Lincoln's intention. Beauregard, the Confederate commander in Charleston and, coincidently Anderson's artillery instructor at West Point, severed Sumter's communications with the mainland and began to marshal forces around the harbor. Three days later the Confederate secretary of war, Leroy Pope Walker, ordered Beauregard to demand the fort's evacuation or bombard it into submission. Anderson's response to the ultimatum delivered on the 11th was to state that he would evacuate in four days time unless he was attacked or received further orders from Washington.

At 0320 hours on the 12th Anderson received a note that read, "We have the honor to notify you that [Beauregard] will open fire in one hour from this time." At 0430 hours Captain George James, commanding an artillery battery on James Island, fired

ABOVE: Kentucky-born West Point graduate Major Robert Anderson, the Union commander of Fort Sumter. After surrendering there in April 1861, he was promoted to brigadier-general and placed in charge of Union forces in Kentucky. Invalided out in October 1863, he finally returned to Sumter on 14 April 1865, to watch the ceremonial raising of the US flag over what little remained of the fort.

RIGHT: Confederate artillery batteries emplaced in Fort Moultrie on Sullivan's Island pour fire into Fort Sumter. With its 45 or so guns under-crewed and lacking ammunition, Sumter's reply was extremely limited. With no hope of succor, the garrison surrendered after thirty-four hours of punishment.

the first shot of the civil war from a 10-inch mortar. Sumter's reply did not come until daylight, but then its fire was intermittent due to a lack of artillerymen and ammunition. The appearance of a number of Union warships briefly gave the garrison hope, but they made no effort to intervene and soon sailed away.

The Confederate fire continued throughout the night and at dawn on the 13th other batteries, especially those at Fort Moultrie, added to the barrage. The fort's barracks were soon in flames and its cannon were reduced to firing just one shot every five minutes or so. A white flag was finally raised in the early afternoon and after some negotiating Anderson agreed to evacuate the next day.

Sumter had been hit by some 4,000 shells in thirty-four hours yet neither side had suffered any deaths. That was soon to change as the garrison stood to attention to salute the Stars and Stripes before it was lowered. Sparks from a smoldering fire ignited a paper-wrapped cannon cartridge and the explosion produced six casualties one of whom, Private Daniel Hough, was the war's first fatality.

ABOVE: Fort Sumter shown before the outbreak of the civil war. It was named after a careerist soldier, Virginia-born General Thomas Sumter, who had served with the British during the French and Indian War (1754–63) and then fought against them in the War of Independence (1775–83). He then settled in South Carolina and entered politics.

RIGHT: The interior of the largely destroyed fort photographed at the end of the conflict. It had been subjected to various attacks by Union forces after 1861. In late 1863, for example, 2,691 shells were lobbed against it and other barrages of similar intensity followed. Yet the fort never succumbed to direct Union pressure, being abandoned only on 17 February 1865.

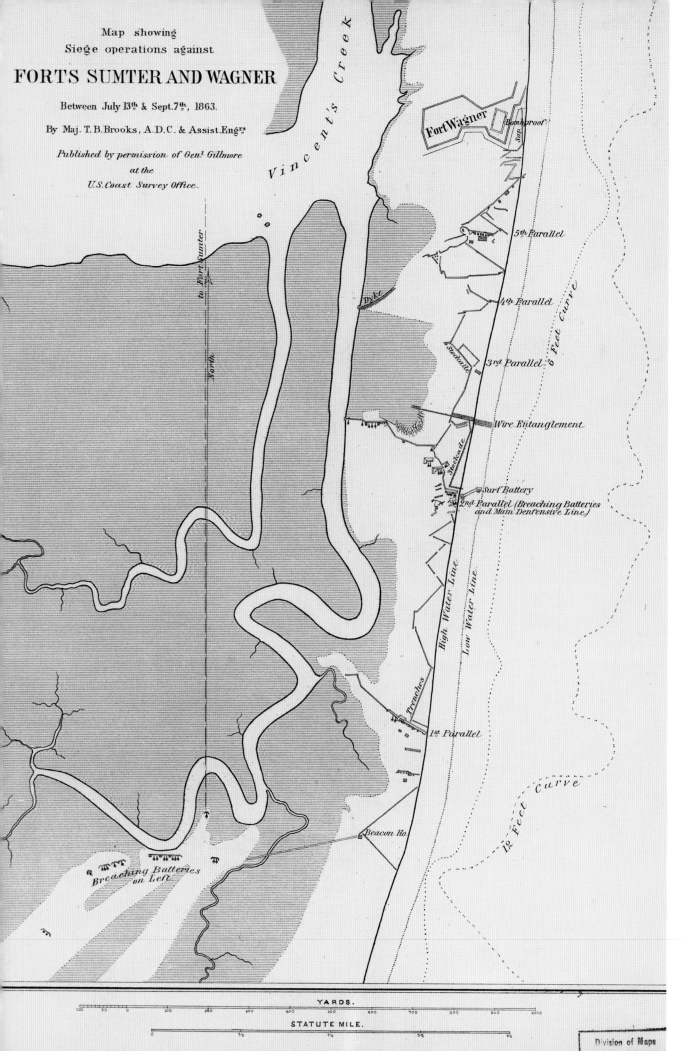

Map showing
Siege operations against

FORTS SUMTER AND WAGNER

Between July 13th & Sept.7th, 1863.

By Maj. T.B.Brooks, A.D.C. & Assist.Engrs.

Published by permission of Genl Gillmore
at the
U.S.Coast Survey Office.

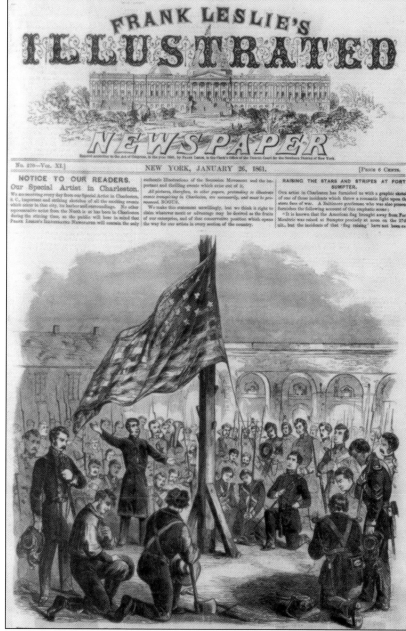

LEFT: Aside from Fort Sumter at the mouth of Charleston Bay, the city was also protected by Forts Moultrie and Fort Wagner on Sullivan's and Morris Islands, respectively. Wagner was attacked by Union land forces on 10 and 18 July 1863. The brigade involved in the first attempt lost 339 men, while the 54th Massachusetts Regiment, an African-American unit with white officers, briefly fought its way into the position during the second attack, only to be evicted with 25 percent casualties.

ABOVE: The garrison of Fort Sumter looks on as Anderson lowers the Stars and Stripes before leaving the fort on 14 April 1861. As this deliberately patriotic and highly emotive illustration suggests, the loss of the fort was used for propaganda purposes to rally support against the Confederacy in the North.

RIGHT: A dramatic depiction of the bombardment of Fort Sumter. While it was certainly true that the position suffered considerable damage from Confederate shells, its garrison did not reply in kind as suggested here.

THE FIRST BATTLE OF BULL RUN, 1861

DATE: 21 July 1861

COMMANDERS: (Union) Brigadier-General Irvin McDowell; (Confederate)
Brigadier-General Joseph Eggleston Johnston

TROOP STRENGTHS: Unon: c.38,000; (Confederate) c. 35,500

CASUALTIES: (Union) 460 killed, 1,124 wounded, and 1,312 missing;
(Confederate) 387 killed, 1,582 wounded, and 13 missing

The Confederacy's leadership knew that its best chance of securing independence was to strike fast and hard against the Union capital, Washington, with its main field force under Beauregard. Equally, the federal authorities believed that the war would be won if they captured the South's capital, Richmond. By July 1861 Beauregard's 20,000 men were located in the vicinity of Manassas, Virginia, protecting a rail line into the Shenandoah Valley, where a second Confederate army, 12,000 troops under Brigadier Joseph E. Johnston, was facing some 18,000 Union troops under Major-General Robert Preston. Johnston was ordered to link up with Beauregard, who was positioned along Bull Run Creek, on the 18th, moving his

RIGHT: The movement and disposition of both Union and Confederate forces during the First Battle of Bull Run, or the First Battle of Manassas as it was called by the Confederates. The Union commander attempted to swing round his opponents' left flank by way of Sudley Springs Ford (top left) from his base at Centreville (centre right), a maneuver that led to bitter fighting around Henry Hill (center left).

FAR RIGHT: An overview of the campaign area in Virginia. The Confederates, who were seeking a decisive victory to bring the war to a swift conclusion, had pushed dangerously close to Washington—Manassas Junction and Centreville are roughly twenty miles to the west of the capital—and thereby made the Union forces under Brigadier-General Irvin McDowell sally forth to meet them.

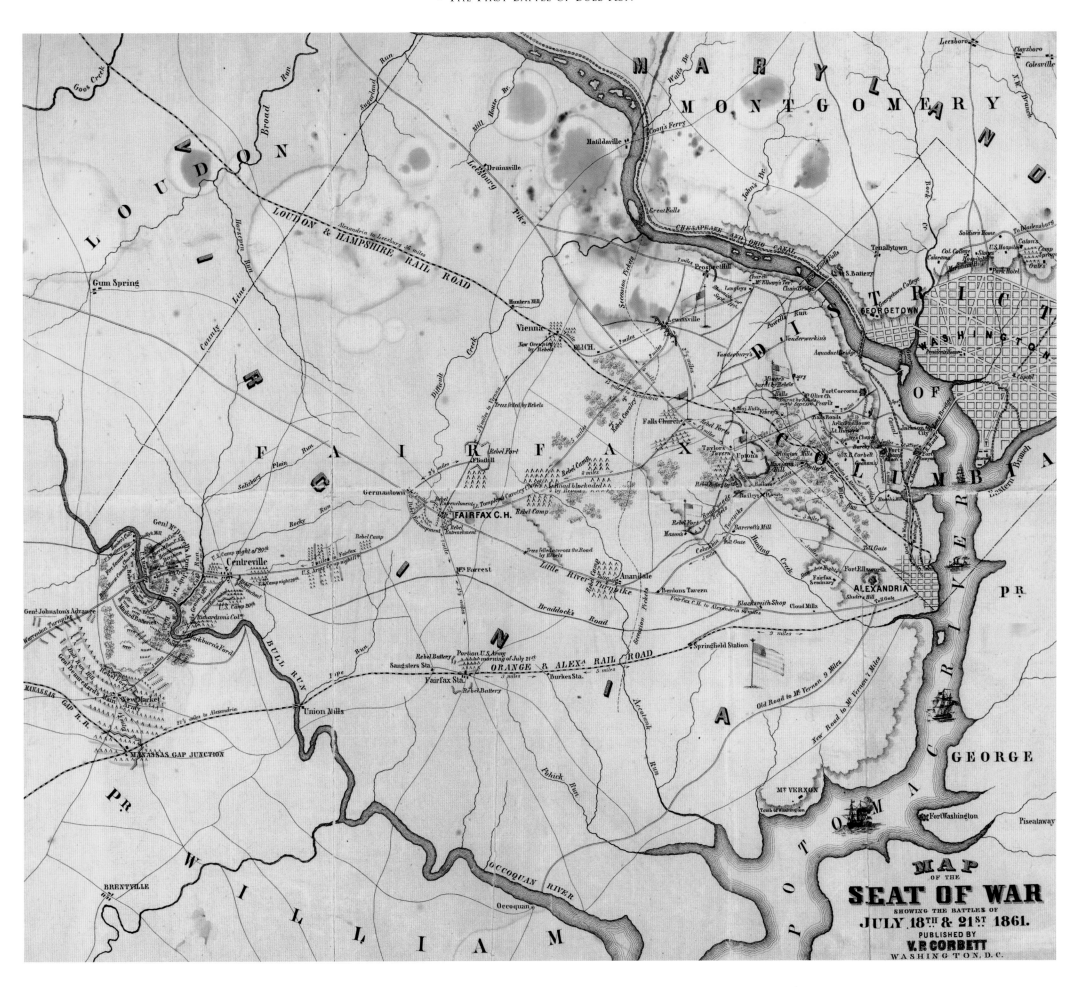

MAP
OF THE
SEAT OF WAR
SHOWING THE BATTLES OF
JULY 18TH & 21ST 1861.
PUBLISHED BY
V. P. CORBETT
WASHINGTON, D.C.

LEFT: The Confederate forces at Bull Run were initially concentrated along the river itself to block various fords and bridges and directly facing their opponents who had concentrated around Centreville. Their commander had to hastily re-orientate his troops when the Union attack developed against his left flank. The critical part of the battlefield was Henry Hill (center left), just south of the Warrenton Turnpike. Union forces pushed south along the Sudley Road (top left), crossed Young's Branch, a tributary of the Bull Run River, and then threw themselves against Henry Hill. Other Union troops crossed the latter river at the Stone Bridge (centre) and marched into action along the aforementioned turnpike.

RIGHT: A closer look at the key sector of the Bull Run battlefield, showing the rival forces at various times during the fighting around Henry Hill (center). The main battle commenced at around 0915 hours and the turning point arrived when two fresh Confederate brigades, some 5000 men under Colonel Jubal Early and Brigadier-General Edmund Kirby Smith, reach Bull Run at 1600 hours. These struck the southern end of the Union line near Bald Hill (bottom left).

entire command by trail in what was a military first. It took just two days to move the bulk of the army some sixty miles. McDowell remained unaware that the transfer had taken place.

McDowell was under acute pressure to march on Richmond but would first have to deal with Beauregard. Bull Run was only really crossable at a small number of fords and one small bridge, so McDowell's options were limited. He decided to move from Centreville and launch a major strike against his opponent's left flank in a maneuver

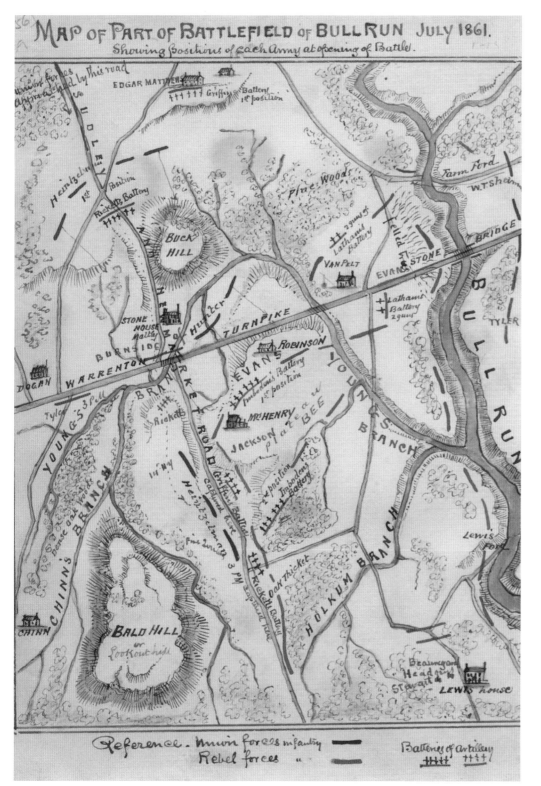

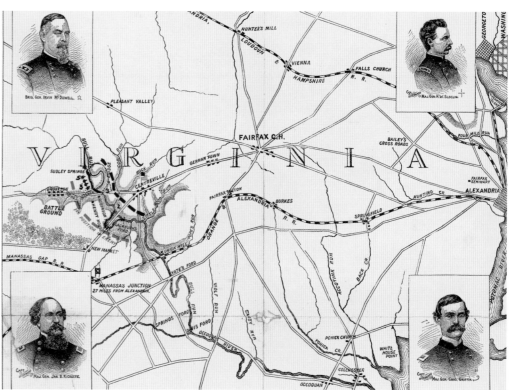

TOP: A Currier & Ives print depicting the collapse of the Union army. Panicking troops flee down the Warrenton Turnpike, crossing the Bull Run River over the Stone Bridge (top left), while the 1st Virginia Cavalry Regiment under Colonel James Ewell Brown "Jeb" Stuart launches a successful charge against the 11th New York Regiment (below right), a zouave unit wearing uniforms mimicking those of French colonial troops from North Africa.

ABOVE: An overview of the Confederate and Union dispositions before the beginning of the battle. The Confederates' line ran along the Bull Run River between Union Mills Ford and the Stone Bridge, but the Union forces made their main push farther to the north.

LEFT: Virginia-born West Point graduate Joseph Eggleston Johnston commanded the Confederate Army of the Shenandoah during the Bull Run campaign and was the ranking officer during the actual battle. He approved the dispositions made by Brigadier-General Pierre Gustave Toutant Beauregard, his second-in-command, and directed the fighting on 21 July. Johnston was nicknamed the "Gamecock" because of his military bearing and jaunty demeanor.

ABOVE: Brigadier-General Thomas Jonathan Jackson was little known before the First Battle of Bull Run but it was where he won his famous "Stonewall" nickname while leading the defense of Henry Hill at a crucial point of the intense fighting along its slopes.

RIGHT: A less than accurate map showing the dispositions of Confederate and Union forces on and around Henry Hill at around 1500 hours on 21 July. Jackson's position is marked by a star and the individual units of his 1st Virginia Brigade, the 2nd, 4th, 5th, 27th, and 33rd Virginia Regiments. These units were officially designated the "Stonewall" Brigade on 30 May 1863.

to isolate Beauregard from Johnston. The attack got underway on the morning of the 21st, but McDowell's opening gambit, a feint attack directed towards Mitchell's Ford Bridge, was spotted for what it really was. His main force crossed over Bull Run at the Stone Bridge and Sudley Springs Ford and then bore down on the Confederate left. The use of signal flags, another battlefield first, alerted Beauregard to the danger and,

just in time, he was able to funnel reinforcements to his exposed left, particularly the area around Henry Hill.

There the battle raged at its most intense and it was there that one Confederate brigade commander, Brigadier-General Thomas Jackson, kept his nerve during the bitter two-hour battle for Henry Hill, earning himself the nickname "Stonewall." The

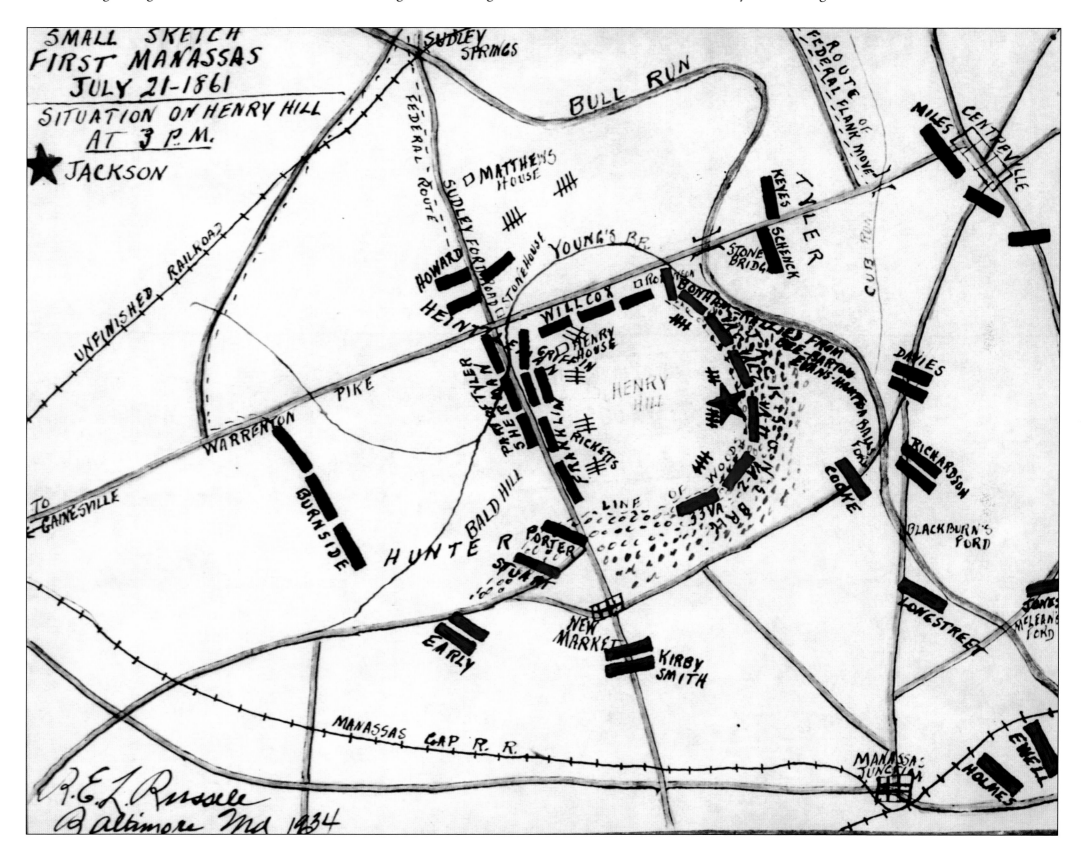

battle turned at around 1600 hours, thanks to the arrival of two fresh Confederate brigades, one of them the last of Johnston's units to reach the battlefield by train. These proceeded to roll up McDowell's right flank. As they retreated under these unexpected hammer blows, some Union units began to panic and simply fled all the way back to Washington. For their part, the Confederates were too exhausted and disorganized to pursue their opponents. The battle fizzled out.

Bull Run was the largest engagement so far fought in North America but it was largely inconclusive, although the South could claim a victory. Washington did not fall, however, and the war would go on. Nevertheless, defeat came as a profound shock to the North, while the narrow victory reinforced the common view among the Confederates that they were martially superior to their opponents. In reality, the performance of the troops on both sides, the vast majority of whom were non-professional, was decidedly mixed.

ABOVE: The Bull Run River was known to be fordable at eight points and crossed by a road bridge at another. But it was in fact fordable elsewhere. One Union commander who was later to gain great fame, Colonel William Tecumseh Sherman, observed a Confederate cavalryman make a crossing at an unmarked ford upstream from the Stone Bridge and he threw his own 3rd Brigade across it at around 1100 hours.

RIGHT: The key features of what is now known as the Manassas National Battlefield Park in Prince William County, north Virginia. Many of the battle's key features have been preserved and various monuments and memorials erected.

FAR RIGHT: A depiction of the rival dispositions at the turning point of the battle, at around 1600 hours. Two fresh Confederate brigades—those of Colonel Jubal Early and Colonel Arnold Elzey, who had taken over from the wounded Brigadier-General Edmund Kirby Smith—are on the far left of the Southern line. These units plus a cavalry regiment under Stuart attacked as if from nowhere and decisively tuned the tide.

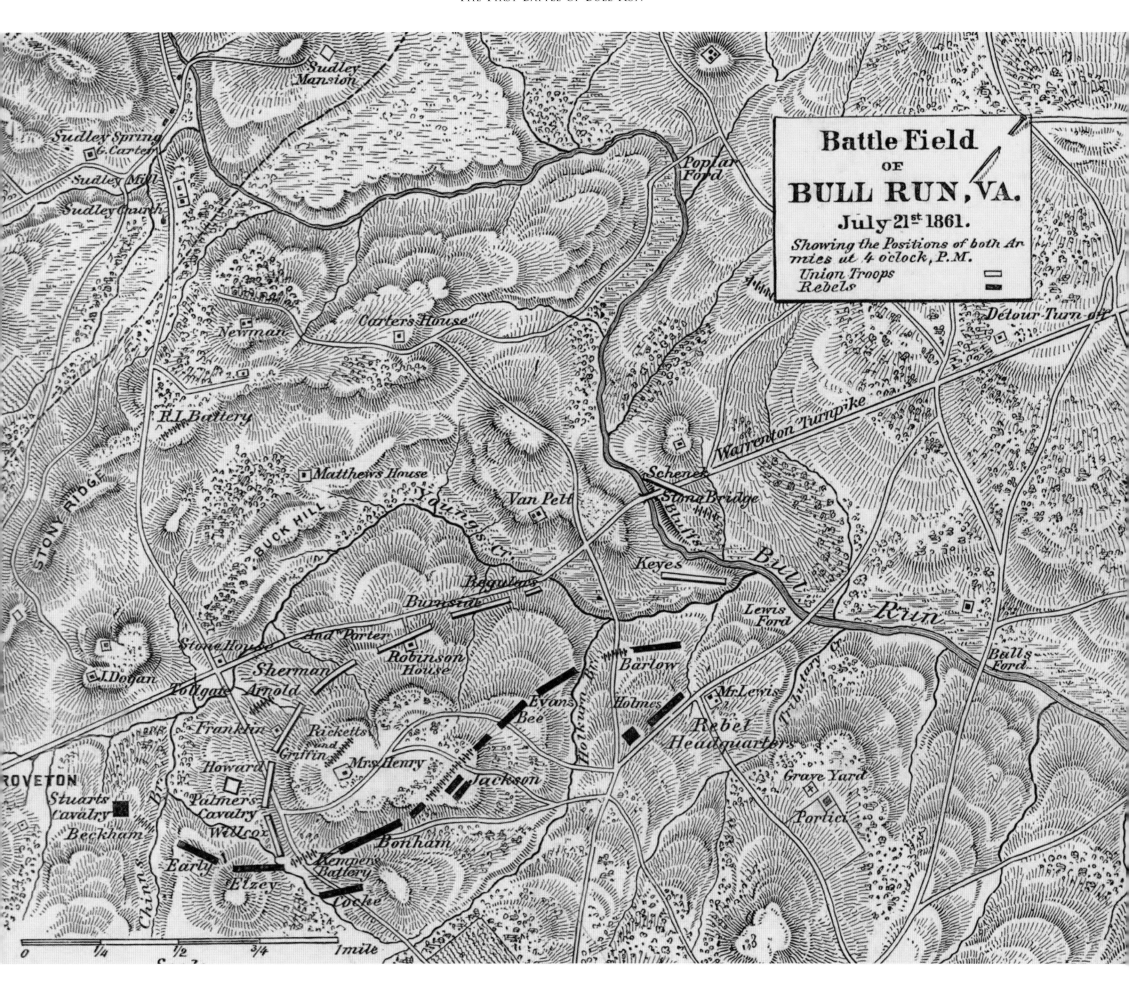

THE BATTLE OF SHILOH, 1862

DATE: 6–7 April 1862

COMMANDERS: (Union) Major-General Ulysses S. Grant; (Confederate) General Albert S. Johnston, Brigadier-General Pierre Gustave Toutant Beauregard

TROOP STRENGTHS: (Union) 62,700; (Confederate) 40,000

CASUALTIES: (Union) 1,754 killed, 8,408 wounded, and 2885 missing; (Confederate) 1,723 killed, 8,012 wounded, and 959 missing

The Confederate cause in Tennessee suffered a number of reverses in February 1862. Grant captured Fort Henry on the Tennessee River on the 6th and went on to secure the unconditional surrender of Fort Donelson on the Cumberland River on the 16th. The state capital, Nashville, fell to Major-General Don Carlos Buell in the last week of the same month. Johnston temporarily abandoned the state, moving some 40,000 men to Corinth just over the border with Mississippi. Grant followed with around a similar number of troops but was ordered to halt on the banks of the Tennessee River around Pittsburg Landing and Shiloh Church, some twenty-two miles north of Corinth, and await the arrival of Buell's troops from Nashville.

Beauregard, who was Johnston's second-in-command with the Army of the

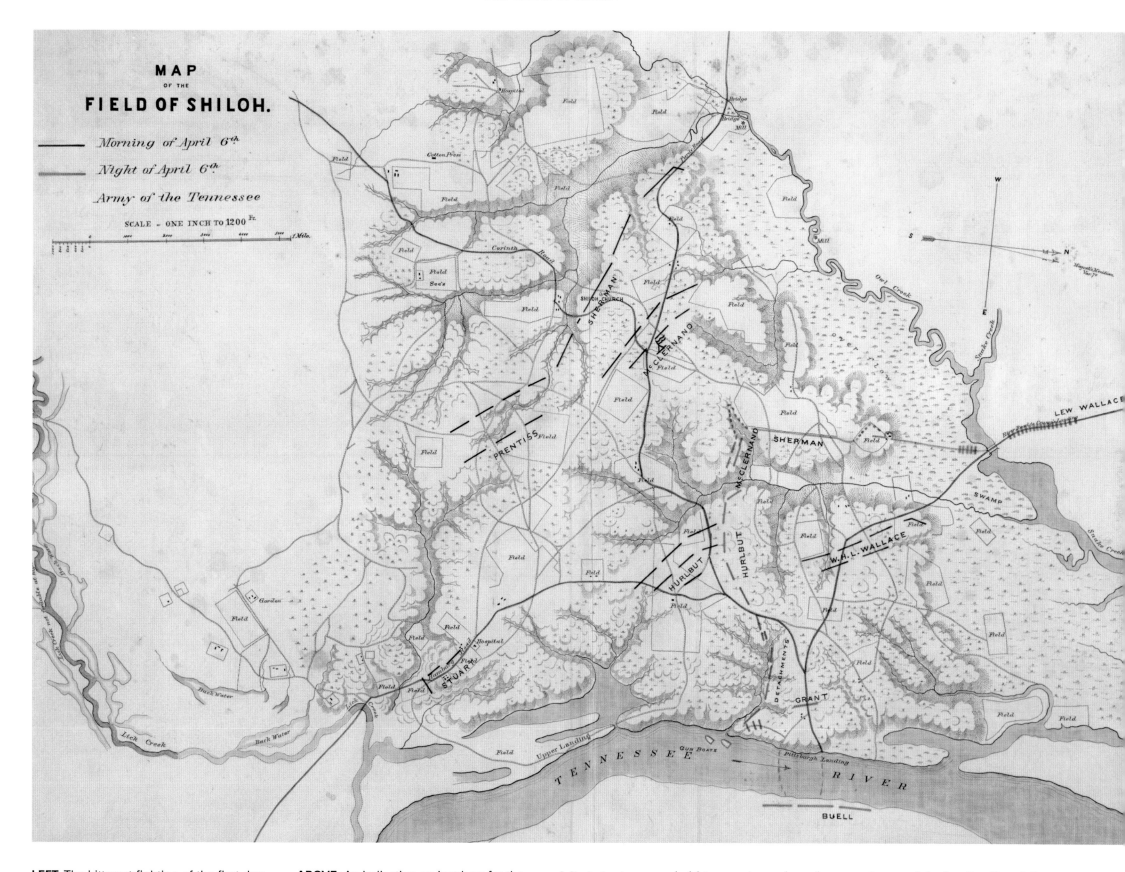

MAP
OF THE
FIELD OF SHILOH.

——————— *Morning of April 6th*

═══════════ *Night of April 6th*

Army of the Tennessee

SCALE - ONE INCH TO 1200 Ft.

LEFT: The bitterest fighting of the first day of the battle took place at a position in thick forest that was dubbed the "Hornet's Nest" by the Confederates.

ABOVE: An indication on just how far the Union line had been pushed back during the first day of the battle. However, Grant's troops were able to reform and more importantly receive reinforcements.

Mississippi, persuaded his superior to launch a surprise attack before Buell and Grant linked up. Johnston agreed, and the move north began on 3 April. Rain, rough terrain, and logistical problems delayed the advance, so that the attack did not begun until the early morning of the 6th. The battle opened before daylight and a succession of forward Union positions were steamrollered by the Confederates. After three hours

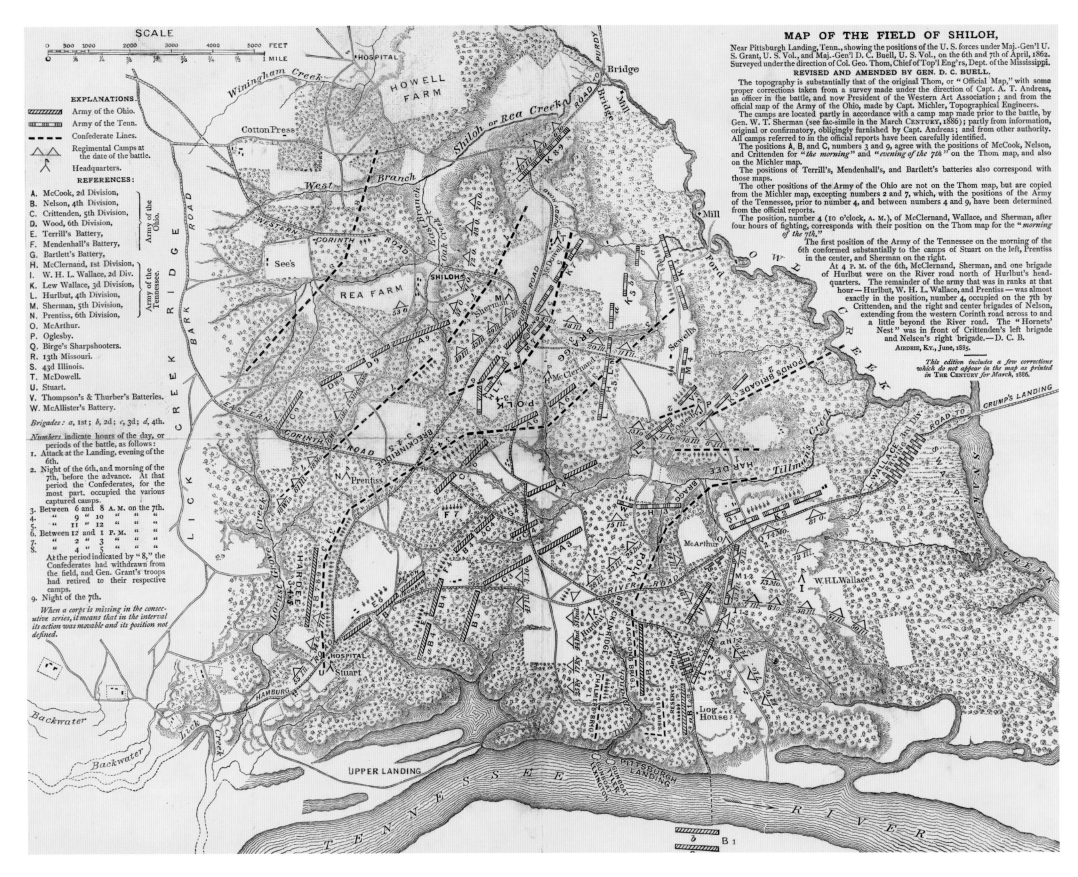

SCALE

0   500   1000   2000   3000   4000   5000 FEET

¼   ½   ⅝   ¾   ⅞   1 MILE

EXPLANATIONS.

Army of the Ohio.
Army of the Tenn.
Confederate Lines.
Regimental Camps at
  the date of the battle.
Headquarters.

REFERENCES:

A. McCook, 2d Division,
B. Nelson, 4th Division,
C. Crittenden, 5th Division,   Army of the
D. Wood, 6th Division,   Ohio.
E. Terrill's Battery,
F. Mendenhall's Battery,
G. Bartlett's Battery,
H. McClernand, 1st Division,
I. W. H. L. Wallace, 2d Div.
K. Lew Wallace, 3d Division,   Army of the
L. Hurlbut, 4th Division,   Tennessee.
M. Sherman, 5th Division,
N. Prentiss, 6th Division,
O. McArthur.
P. Oglesby.
Q. Birge's Sharpshooters.
R. 13th Missouri.
S. 43d Illinois.
T. McDowell.
U. Stuart.
V. Thompson's & Thurber's Batteries.
W. McAllister's Battery.

Brigades: a, 1st; b, 2d; c, 3d; d, 4th.

Numbers indicate hours of the day, or
periods of the battle, as follows:
1. Attack at the Landing, evening of the
   6th.
2. Night of the 6th, and morning of the
   7th, before the advance. At that
   period the Confederates, for the
   most part. occupied the various
   captured camps.
3. Between 6 and 8 A. M. on the 7th.
4. "   "   9 " 10 "   "   "
5. "   " 11 " 12 "   "   "
6. Between 12 and 1 P. M.
7. "   "   2 "   3 "   "   "
8. "   "   4 "   "   "
At the period indicated by "8," the
   Confederates had withdrawn from
   the field, and Gen. Grant's troops
   had retired to their respective
   camps.
9. Night of the 7th.

*When a corps is missing in the consec-
utive series, it means that in the interval
its action was movable and its position not
defined.*

MAP OF THE FIELD OF SHILOH,

Near Pittsburgh Landing, Tenn., showing the positions of the U. S. forces under Maj.-Gen'l U.
S. Grant, U. S. Vol., and Maj.-Gen'l D. C. Buell, U. S. Vol., on the 6th and 7th of April, 1862.
Surveyed under the direction of Col. Geo. Thom, Chief of Top'l Eng'rs, Dept. of the Mississippi.

REVISED AND AMENDED BY GEN. D. C. BUELL.

The topography is substantially that of the original Thom, or "Official Map," with some
proper corrections taken from a survey made under the direction of Capt. A. T. Andreas,
an officer in the battle, and now President of the Western Art Association; and from the
official map of the Army of the Ohio, made by Capt. Michler, Topographical Engineers.
  The camps are located partly in accordance with a camp map made prior to the battle, by
Gen. W. T. Sherman (see fac-simile in the March CENTURY, 1886); partly from information,
original or confirmatory, obligingly furnished by Capt. Andreas; and from other authority.
All camps referred to in the official reports have been carefully identified.
  The positions A, B, and C, numbers 3 and 9, agree with the positions of McCook, Nelson,
and Crittenden for "the morning" and "evening of the 7th" on the Thom map, and also
on the Michler map.
  The positions of Terrill's, Mendenhall's, and Bartlett's batteries also correspond with
those maps.
  The other positions of the Army of the Ohio are not on the Thom map, but are copied
from the Michler map, excepting numbers 2 and 7, which, with the positions of the Army
of the Tennessee, prior to number 4, and between numbers 4 and 9, have been determined
from the official reports.
  The position, number 4 (10 o'clock, A. M.), of McClernand, Wallace, and Sherman, after
four hours of fighting, corresponds with their position on the Thom map for the "morning
of the 7th."
  The first position of the Army of the Tennessee on the morning of the
6th conformed substantially to the camps of Stuart on the left, Prentiss
in the center, and Sherman on the right.
  At 4 P. M. of the 6th, McClernand, Sherman, and one brigade
of Hurlbut were on the River road north of Hurlbut's headquarters. The remainder of the army that was in ranks at that
hour—Hurlbut, W. H. L. Wallace, and Prentiss—was almost
exactly in the position, number 4, occupied on the 7th by
Crittenden, and the right and center brigades of Nelson,
extending from the western Corinth road across to and
a little beyond the River road. The "Hornets'
Nest" was in front of Crittenden's left brigade
and Nelson's right brigade.—D. C. B.

AIRDRIE, KY., June, 1885.

*This edition includes a few corrections
which do not appear in the map as printed
in THE CENTURY for March, 1886.*

of fighting, Johnston's leading units had pushed the Union right flank back nearly a
mile.

The Confederates now concentrated their efforts on the Union center but at least
a dozen attacks against a stoutly defended area known as the Hornet's Nest were
thrown back. The fighting in the center also precluded the Confederates from
launching their intended strike against Grant's left flank. Matters were made worse at
1440 hours when Johnston died after having an artery in his leg severed by a stray
bullet. Beauregard took his place. The only good news came at around three hours

LEFT: A highly complex map showing the location of the rival forces at various stages during the two-day Battle of Shiloh. Produced after the war, it pays scant attention to the precise naming or disposition of the Confederate Army of the Mississippi but goes into great detail of the deployment of the Union's Army of the Ohio and Army of the Tennessee.

ABOVE: Major-General Don Carlos Buell commanded the Union's Army of the Tennessee and was able to rush some of his divisions to Shiloh on the first day.

RIGHT: General Albert Sidney Johnston initiated the Battle of Shiloh but was mortally wounded on the first day while leading a charge that cleared a ten-acre peach orchard on the flank of the infamous Hornet's Nest.

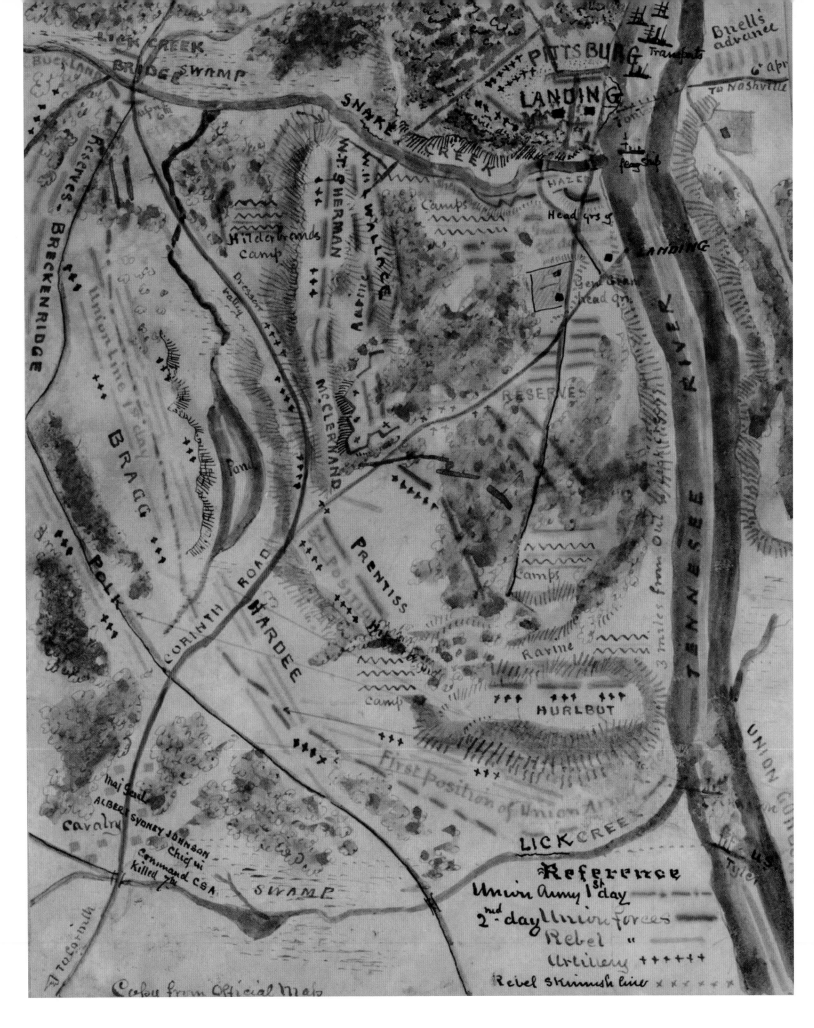

LEFT: The rival dispositions of both Union and Confederate forces during the two days of fighting at Shiloh. The Union line was pushed back a considerable distance on 6 April but Grant's counterattack the next day, which began at around 0730 hours, recaptured virtually all of the lost ground in just three hours. The Confederate commander ordered a full retreat at 1500 hours.

RIGHT: Grant, chewing on his usual cigar, directs the Union counterattack on the second day of the Battle of Shiloh. This image gives the general a traditionally heroic and dashing look but in reality his demeanor and dress were considerably different. During the battle, for example, the general was actually on crutches after a riding accident. Although Shiloh was his victory, it was exceptionally costly— roughly 25 percent of his men had been killed, wounded or captured—and it briefly put his career in jeopardy.

later when the Hornet's Nest was finally taken, but the fighting then ebbed away with sunset.

The Confederates had received intelligence that Buell was near Decatur, northern Alabama, but, in fact, his troops had begun crossing the Tennessee River during the late afternoon of the first day of the battle. One of Grant's own divisions had also arrived at Shiloh and, with these reinforcements more than making up for his losses, he went over to the attack on the 7th. Beauregard, who had no reinforcements to call

on, ordered his troops to stand fast, and for much of the day they held off the Union forces. By the mid-afternoon it was clear to Beauregard that he was risking wholesale defeat if he stayed at Shiloh and so ordered his army back to Corinth.

Both sides claim Shiloh as a victory, but Grant had the better case. He retained possession of the battlefield, the greater part of Tennessee remained under Union control, and he and Buell had linked up. In late May the Confederates abandoned Corinth, and the greater part of western Tennessee, including Memphis, were thus lost.

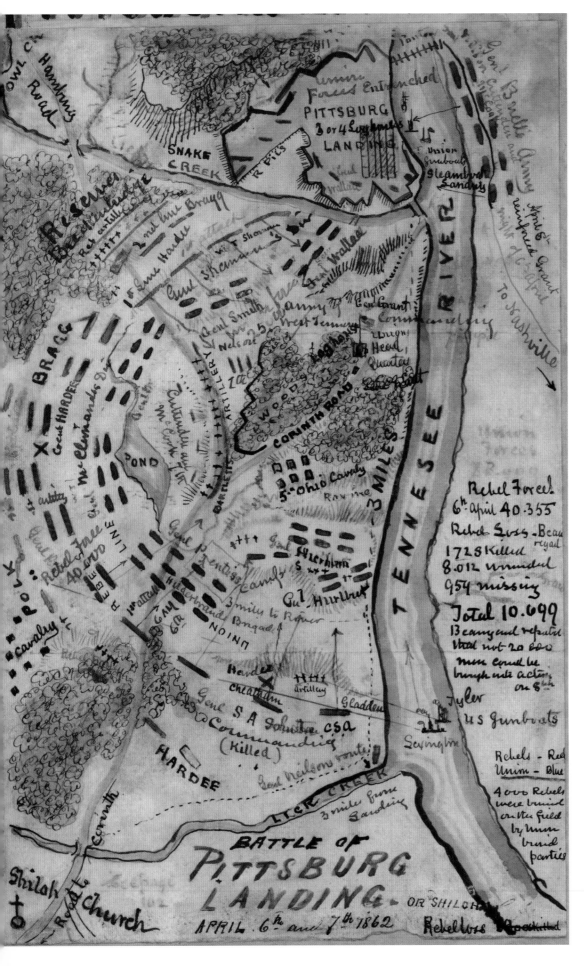

LEFT: A somewhat crude map of the first day of the battle, showing the main Confederate attacks on left and right flanks of Grant's command, aimed at pushing the Union left flank away from the river.

BELOW: Major-General Braxton Bragg had command of the Confederate Army of the Mississippi's II Corps during the battle. On the first day the four Confederate corps were arrayed one behind the other in lines.

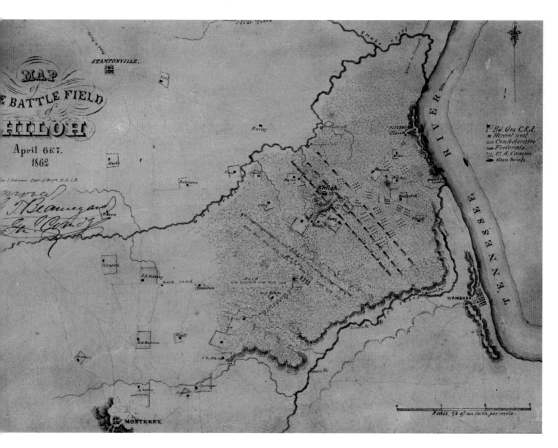

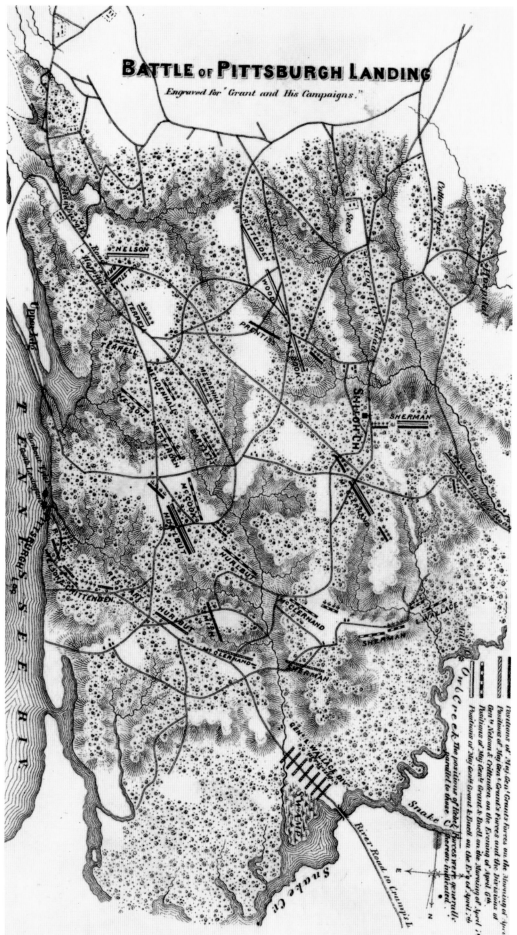

TOP: This map is not completely accurate, but it does indicate the broken, wooden nature of the terrain over which the bulk of the fighting took place. Grant's predicament is plain to see, however. If forced into headlong retreat, his troops would have found escape across the Tennessee River extremely difficult.

ABOVE: The fighting at Shiloh was mostly at close range and casualties were severe on both sides with around 24,000 men in all becoming casualties.

RIGHT: A map detailing the location of various Union divisions over the two days of the Battle of Shiloh.

THE SECOND BATTLE OF BULL RUN, 1862

DATE: 29–30 August 1862

COMMANDERS: (Union) Major-General John Pope; (Confederate) General Robert E. Lee

TROOP STRENGTHS: (Union) 75,000; (Confederate) 54,268

CASUALTIES: (Union) 14,449 killed, wounded or missing; (Confederate) 9,420 killed, wounded or missing

Pope had taken command of the Union's Army of Virginia on 26 June 1862, and over the next few weeks began to plan a campaign to capture the South's capital, Richmond. His 66,000 men were to be supplemented by elements of the Army of the Potomac, which had suffered a mauling outside the Southern capital during the Seven Days' Battles fought between 25 June and 1 July. The Army of the Potomac received orders to move on 3 August. If these two Union forces had been united they would have a combined strength of more than 100,000 men. The main Confederate field force, Lee's Army of Northern Virginia, which was facing Pope along the

RIGHT: The Second Battle of Bull Run, which was known in the South as the Second Battle of Manassas, was brought about largely to destroy one Union force, Major-General John Pope's Army of Virginia, before it could be united with a second, the Army of the Potomac under Major-General George McClellan. The chosen battlefield was close to that of the first Battle of Bull Run and, by a strange coincidence, once again the Confederates held the high ground and their desperate situation in the latter stages of the fighting was transformed by the timely arrival of reinforcements.

FAR RIGHT: This map shows the buildup of the battle from the Union point of view, with troops beginning to mass around Centreville, which is located on a piece of high ground to the north of Blackburn's Ford over the Bull Run River (center). The site of the first battle (center left) on the west bank of the river is visible,

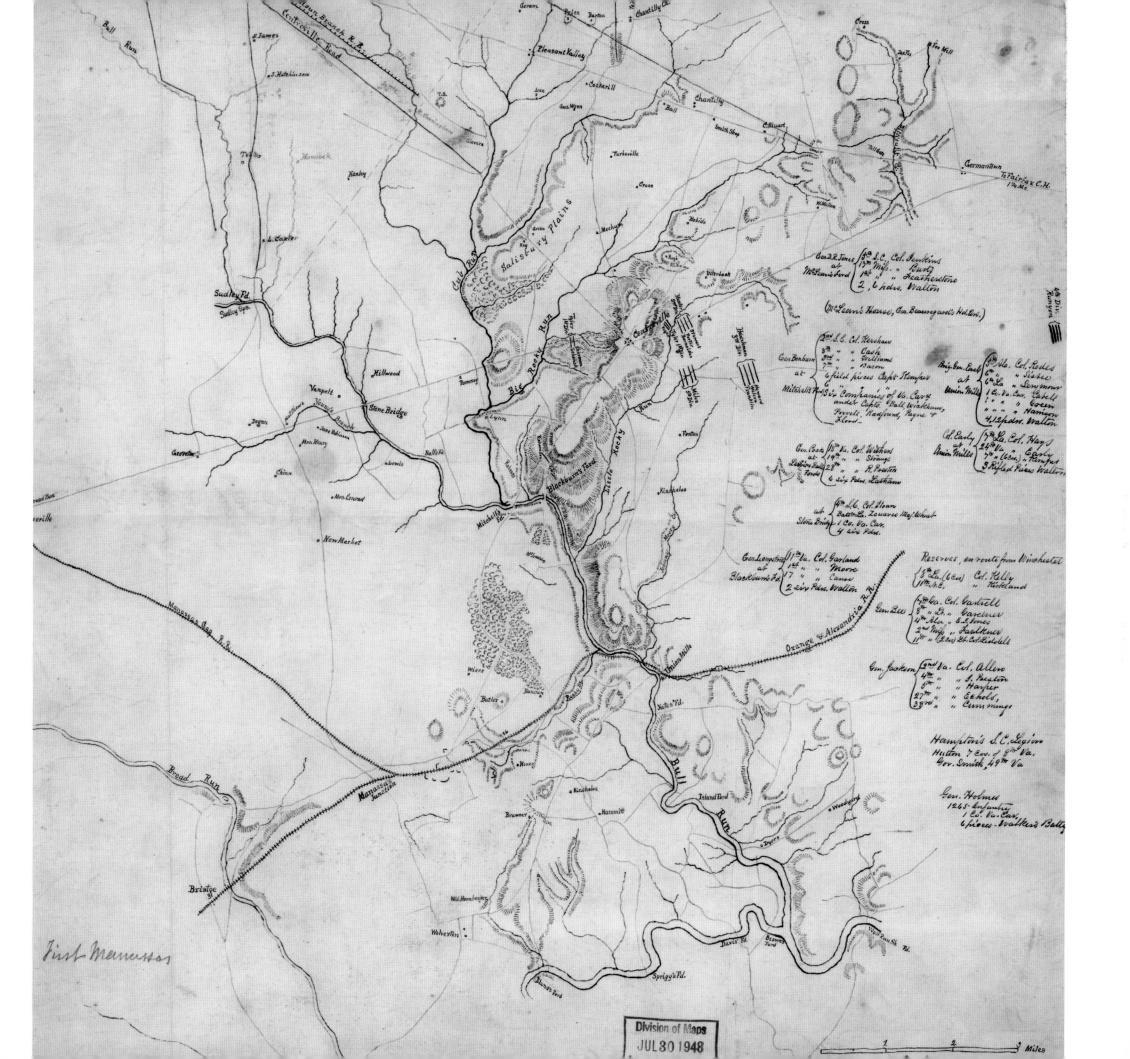

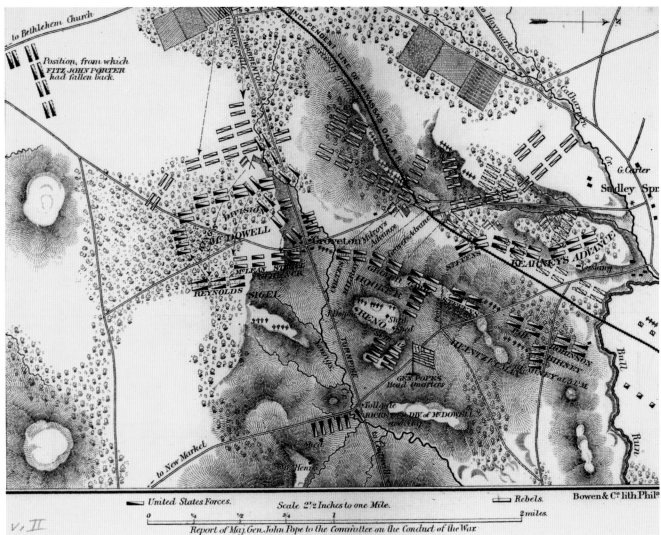

BELOW LEFT: This map puts Major-General Fitz-John Porter's command (top left) wildly out of position. In the actual fighting it was first blocked along a river known as Dawkin's Branch.

LEFT: Although far from reality, not least for the pristine nature of the uniforms, this illustration gives a fair impression on the intensity of the fighting during the battle.

BELOW: Lieutenant-General James Longstreet, undoubtedly one of the finest Confederate tactical commanders of the war but one who showed a less sure touch when it came to strategic matters. "Old Pete" was highly popular with the men he led into battle.

No. 2.
ILLUSTRATIVE
MAP
OF THE
BATTLE-FIELD OF MANASSAS, VA.
SHOWING POSITIONS AND MOVEMENTS OF TROOPS
August 29ᵗʰ, 1862.
TO ACCOMPANY CLOSING ARGUMENT
OF
COUNSEL FOR THE GOVERNMENT
POSITIONS LAID DOWN BY HIM
TIME OF DAY 6, P.M.
SCALE

Rappahannock River in central Virginia, comprised a mere 55,000 men.

When Lee received news of the Army of the Potomac's movement on 24 August, he immediately realized that he had to strike first before the two Union armies could unite against him. That same day the Confederate commander ordered Major-General Thomas "Stonewall" Jackson's 12,000 men to march north and then east to get behind Pope's men in the vicinity of Manassas. Lee would follow with Lieutenant-General James Longstreet's corps. Jackson marched an epic fifty-four miles in high summer, reaching the vicinity of Manassas on the 27th. The next day he looted and burned Pope's supply dump at Manassas Junction and then took up a defensive position along the line of an incomplete railway cutting a little to the west of the First Bull Run battlefield of the previous year.

ABOVE: A splendid map showing the general dispositions of both sides during the battle. Of greatest significance is the red line running from Thoroughfare Gap (top left) to the southwest edge of the battle by way of Gainesville.

RIGHT: The interior of a Union fort. Many such defenses were thrown up around Washington, particularly after the defeat during the First Battle of Bull Run..

To further lure the Union commander in the trap. Jackson also deliberately revealed his precise location on the 28th by attacking a Union division at nearby Groveton. Pope reacted quickly and concentrated against Jackson the next day. Federal units launched several piecemeal and uncoordinated attacks against the Confederate line, but all were repulsed. Pope resolved to continue the fight the next day but, unbeknownst to him, Longstreet had nearly reached Jackson. The fighting on the 30th saw the Federal troops again assault Jackson's line. None of the attacks was successful, but Jackson's men were fast running out of ammunition and, indeed, some were forced to fling rocks at the Union troops.

Salvation came when Longstreet's command began its large-scale onslaught against the lightly defended and exposed left flank of Pope's force. The Union line was rolled up and thrown back across Bull Run Creek. But, unlike the conclusion of the first battle, this time the withdrawal back to heavily defended Washington was conducted in generally good order. Lee ordered a follow-up the next day but called off the pursuit after Jackson was checked at the Battle of Chantilly on 1 September. Pope was relieved of command eleven days later and the Army of Virginia was absorbed into the Army of the Potomac.

ABOVE: Civil War troops on the move. This somewhat leisurely sketch hardly does justice to the speed and drive that allowed Jackson's men and other Confederate forces to undertake long-distance speed marches.

RIGHT: The Union Army of Virginia enjoyed a sizeable numerical advantage over Jackson's force, and orders were given to concentrate the widely scattered Federal divisions against the Rebel force. The various Union units arrived piecemeal, apart from Major-General Fitz-John Porter's corps, which could not reach its intended position at Gainesville (center right) as it was blocked by a small Confederate force along Dawkin's Creek.

FAR RIGHT: The first day of the battle with the dispositions during the late afternoon on 29 August. One of Pope's corps is heavy engaged against Jackson along the line of an unfinished railroad near the ridge of the Sudley Mountains (top left).

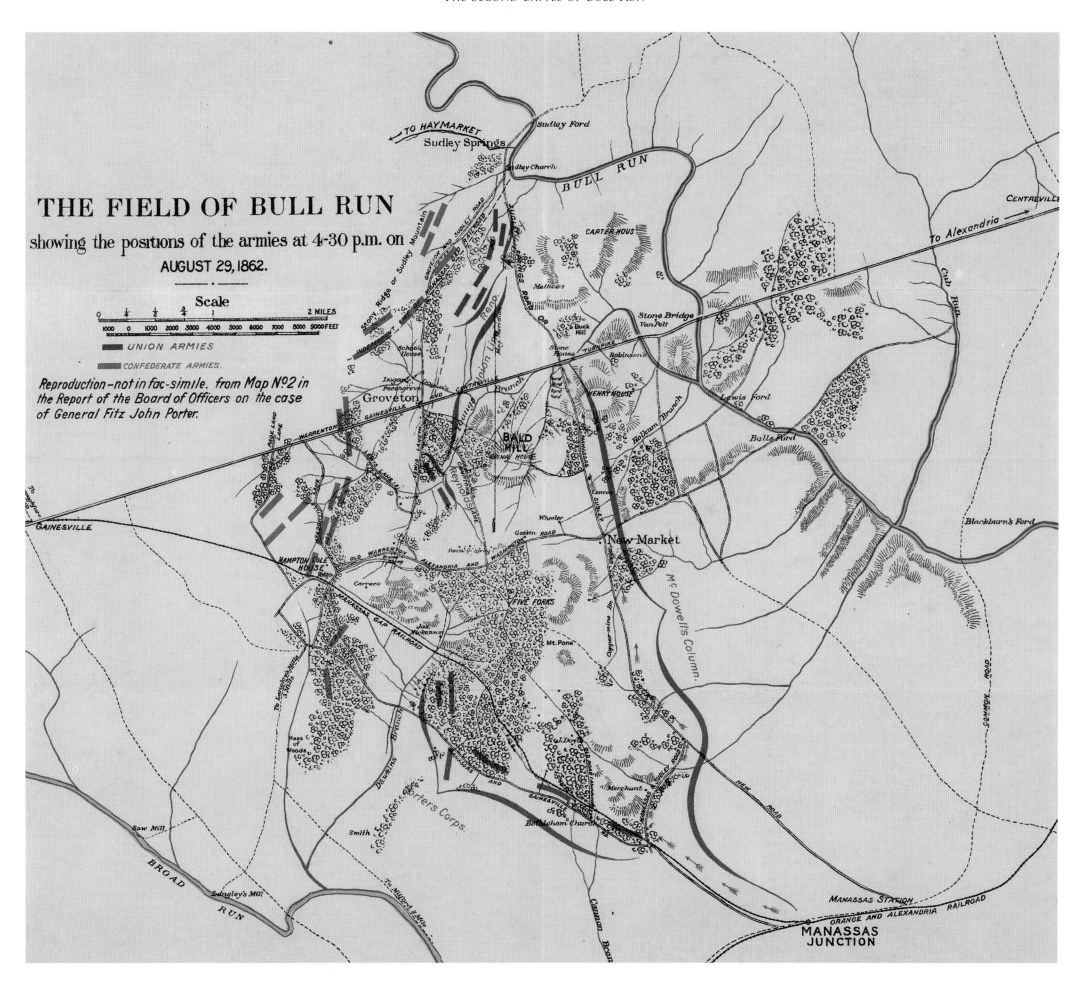

THE FIELD OF BULL RUN

showing the positions of the armies at 4·30 p.m. on
AUGUST 29, 1862.

Scale

1000   0   1000 2000 3000 4000 5000 6000 7000 8000 9000 FEET

UNION ARMIES
CONFEDERATE ARMIES.

Reproduction–not in fac-simile. from Map Nº 2 in
the Report of the Board of Officers on the case
of General Fitz John Porter.

THE BATTLE OF ANTIETAM, 1862

DATE: 17 September 1862

COMMANDER: (Union) Major-General George B. McClellan; (Confederate) General Robert E. Lee

TROOP STRENGTHS: (Union) 80,000; (Confederate) 45,000

CASUALTIES: (Union) 2,108 killed and 10,032 wounded or missing; (Confederate) 2,700 killed and 11,023 wounded or missing

The first six months of 1862 had seen the Confederacy suffer a number of reverses in the various theaters of the civil war. Union forces had overrun something like a total of 50,000 square miles of its territory and the large Army of the Potomac under McClellan was advancing up the Virginia Peninsula to menace Richmond, capital of the South. Yet, within the space of a few months, the situation was utterly transformed, with the attack on the capital defeated and with Confederate troops within easy striking distance of Washington. The catalyst of change was Robert E.

Lee, who on 1 June was given command of the newly created Army of Northern Virginia.

After victory at the Second Battle of Bull Run in late August, Lee paused briefly and then launched an invasion of the North, with his troops beginning to cross the Potomac River on 4 September. McClellan, only recently appointed to command of

LEFT: A contemporary photograph of the so-called "Burnside Bridge" on the southern flank of the Antietam battlefield. The local Union commander, Major-General Ambrose Everett Burnside, spent several hours on the morning of 17 September trying to push his troops over the bridge.

ABOVE: A Currier & Ives lithograph depicting the fierce fighting around "Burnside Bridge" does not really convey that just two Confederate regiments held up an entire Union corps for several hours before being forced to retreat at around 1300 hours.

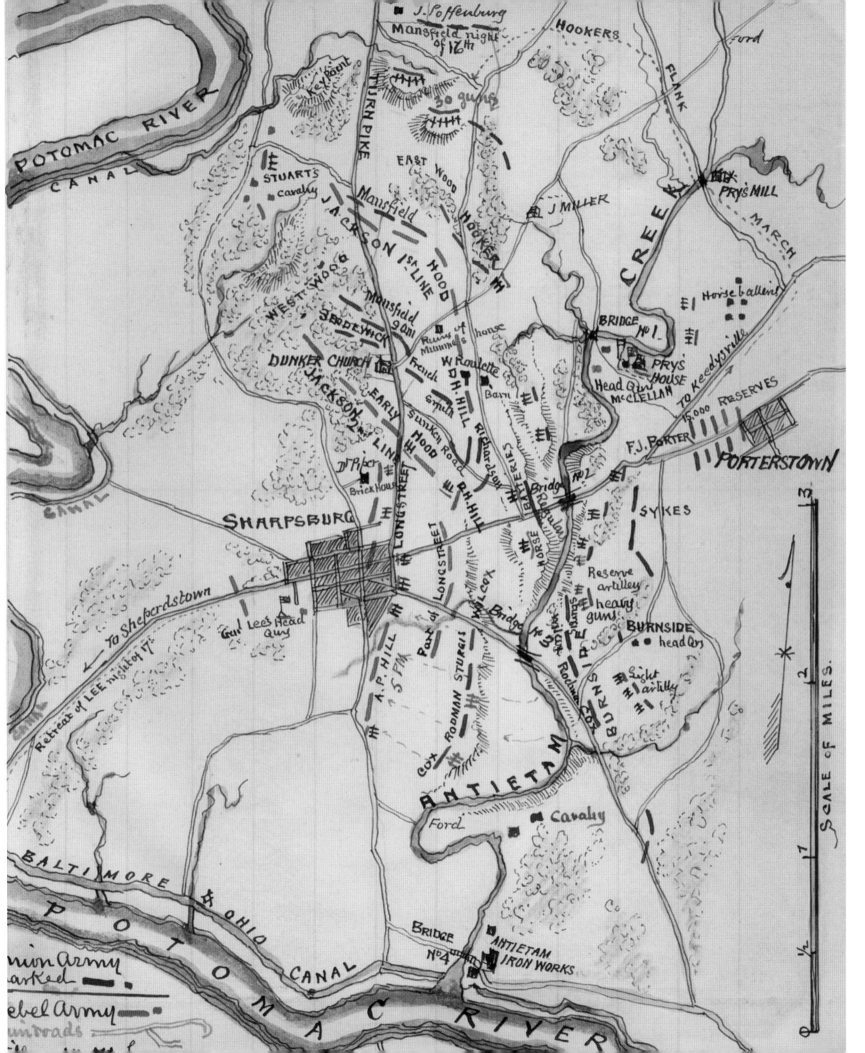

LEFT: A largely accurate map of Antietam, showing the Confederate line to the north and east of Sharpsburg, their name for the battle, and the Union forces lined up each side of Antietam Creek. Many of the battle's key features and events are also shown, including Burnside Bridge, here designated Bridge No. 3, southeast of Sharpsburg, Dunker Church to the north of the town, which Union forces finally captured at around 0900 hours, and to the latter's southeast a sunken road, which was nicknamed "Bloody Lane" for the fierceness of the fighting around it.

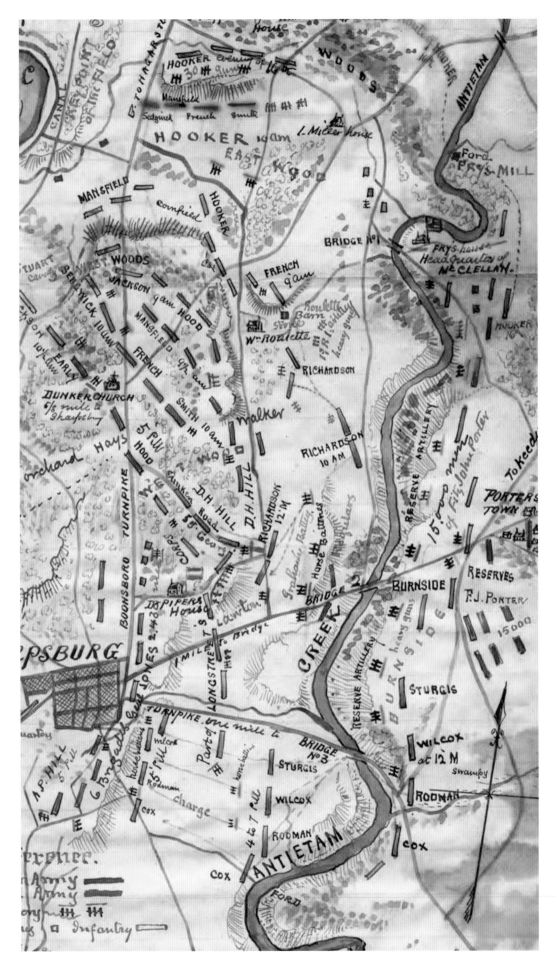

LEFT: A map showing various Union and Confederate dispositions at different stages of the battle. The Rebels were gradually pushed back over a long day's fighting but were saved by the arrival of a fresh division, that of Major-General Ambrose Powell Hill, on their right flank.

BELOW: The main danger faced by the Confederates at Antietam was that their right flank might be pushed back through Sharpsburg, which would cut off the Army of Northern Virginia from its only viable line of retreat over the Potomac River back into friendly territory.

ABOVE: Despite his sluggish, dithering performance during the battle, Burnside was given command of the Union's Army of the Potomac the following November only to be sacked in early 1863.

the Army of the Potomac, moved northwards along the same river, looking for Lee. Two of the Union commander's troops had the good fortune to discover a copy of Lee's invasion orders at Frederick, Maryland, on the 14th. These showed that, at that time, Lee's Army of Northern Virginia was divided into five parts to carry out various tasks. Yet McClellan continued to move slowly, thereby allowing Lee to largely reunited his scattered forces and take up an excellent defensive position on high ground to the east of the town of Sharpsburg. His left flank rested on the Potomac, while to his front ran Antietam Creek.

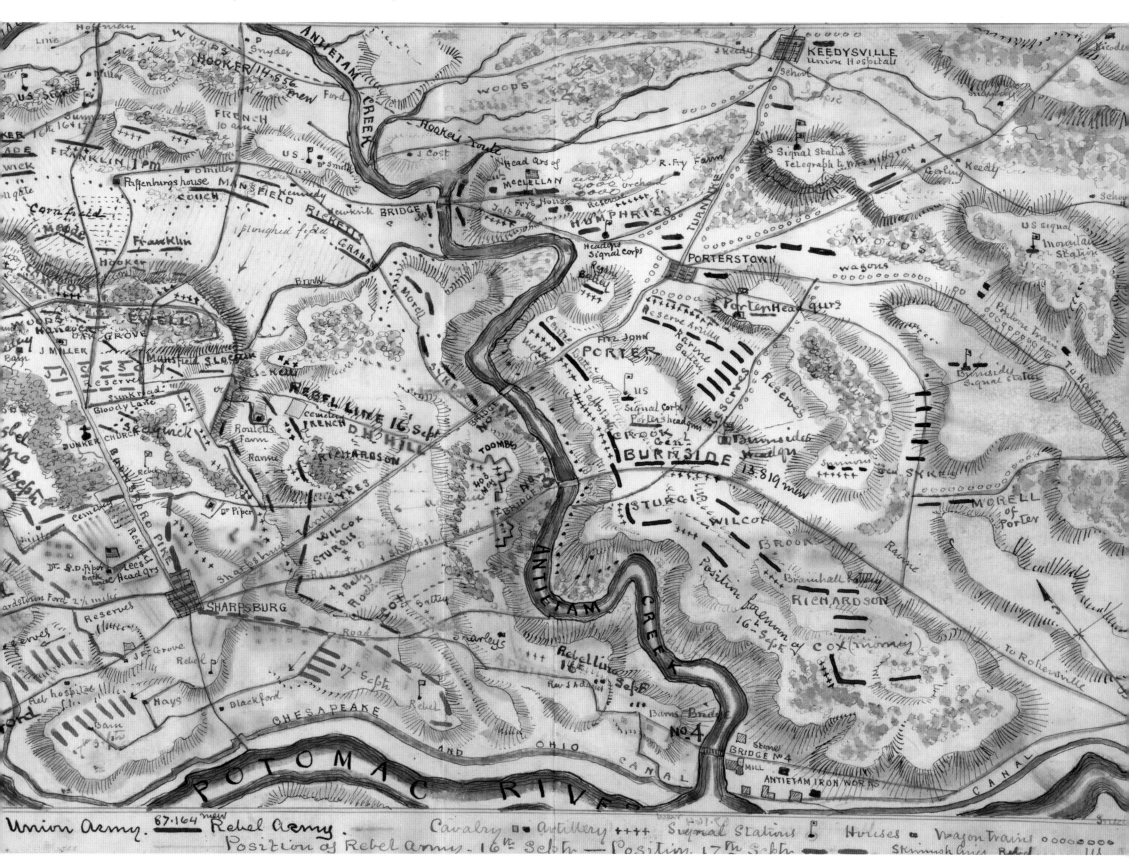

McClellan made the assumption that he was outnumbered but finally attacked on the 17th. He planned to strike at Lee's left flank and then punch across the creek on the right. The early morning attack on the left flank was brought to a halt with heavy casualties, not least along a sunken road later nicknamed "Bloody Lane." The action switched to Lee's center and right during the mid-morning. Union troops made three desperate attempts to take a narrow bridge, not realizing that the creek was fordable. Finally carried, it took the local commander some two hours to reorganize his various units before they began to push towards Sharpsburg.

Yet, even as they neared the town, they were hit on the left flank when a fresh Confederate division arrived from Harpers Ferry. The Union line collapsed and fled back across the creek, thus ending the bloodiest day in US military history. Lee had won a great tactical victory but began withdrawing back across the Potomac two days later. As his invasion was therefore over, the Union could claim the greater strategic victory. Antietam's even greater political significance was revealed on the 22nd when Lincoln felt confident enough to announce his provisional Emancipation Proclamation to free the South's slaves.

LEFT: Union forces under Burnside made increasingly desperate attempts to cross the creek by way of a narrow bridge (the Rorhbach Bridge, subsequently renamed Burnside Bridge) literally squandering his men under ferocious fire, whereas less than a mile downstream the creek was easily fordable, with much lighter opposition.

RIGHT: An overhead view of the Antietam battlefield shows that the Confederate Army of Northern Virginia was in real danger of being trapped against the Potomac River if its lines had wholly given way during the battle. As it was, its divisions were forced back some distance but the Union forces were never able to overwhelm them entirely. Nevertheless, it was a victory of sorts for the Federals and allowed Lincoln to make his Emancipation Proclamation.

THE BATTLE OF STONES RIVER, 1862-63

DATE: 31 December 1862–3 January 1863

COMMANDERS: (Union) Major-General William Rosecrans; (Confederate) General Braxton Bragg

TROOP STRENGTHS: (Union) 44,000; (Confederate) 38,000

CASUALTIES: (Union) 12,906 killed, wounded or missing; (Confederate) 11,740 killed, wounded or missing

The campaigning across Tennessee and Kentucky in the latter part of 1862 had seen Bragg's reinforced Army of Tennessee score a minor but important victory at Richmond, Kentucky, on 30 August, one that led Union Major-General Don Carlos Buell's Army of the Ohio to retreat to the Ohio River, thereby abandoning much of the state. Union reinforcements were sent and advanced back into Kentucky. A drawn battle took place at Perryville on 8 October and, somewhat unnecessarily, Bragg opted to fall back into Tennessee and take up positions around Murfreesboro to protect the railway line that ran from Nashville to Georgia. The tardy Buell was dismissed some two weeks later and replaced by Rosecrans, who reoccupied Nashville

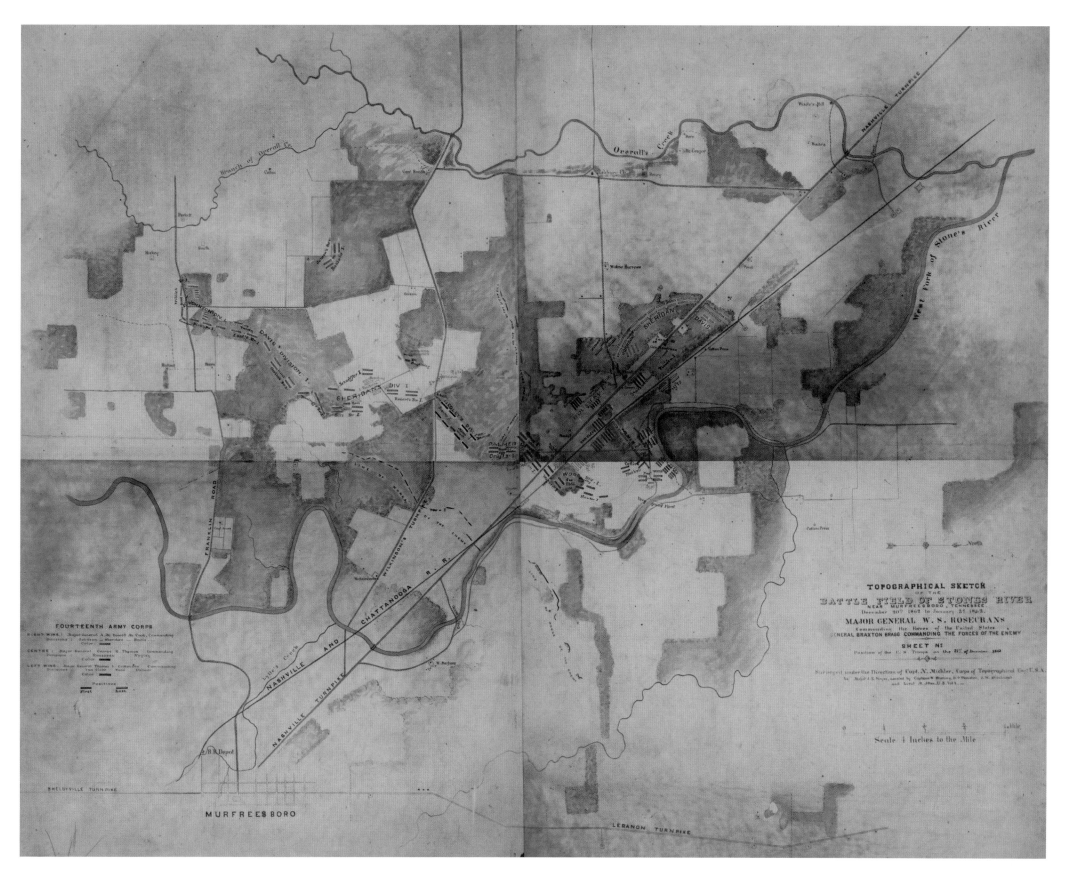

LEFT: Troops fight to cross Stones River during the most costly battle to take place in the civil war's western theater. In reality, much of the major fighting was done away from the river on the Confederate right (Union left) flank. The former even withdrew four out of five of the brigades of Major-General John Cabell Breckinridge's command from their own right flank to the left and away from the river.

ABOVE: The Stones River battlefield, showing the relative positions of the opposing forces. The Confederates placed the greater part of their forces on the western side of the river, where they made their greatest effort on 31 December, a long day of often bitter fighting in which both sides exhausted themselves but were unable to deliver a decisive blow.

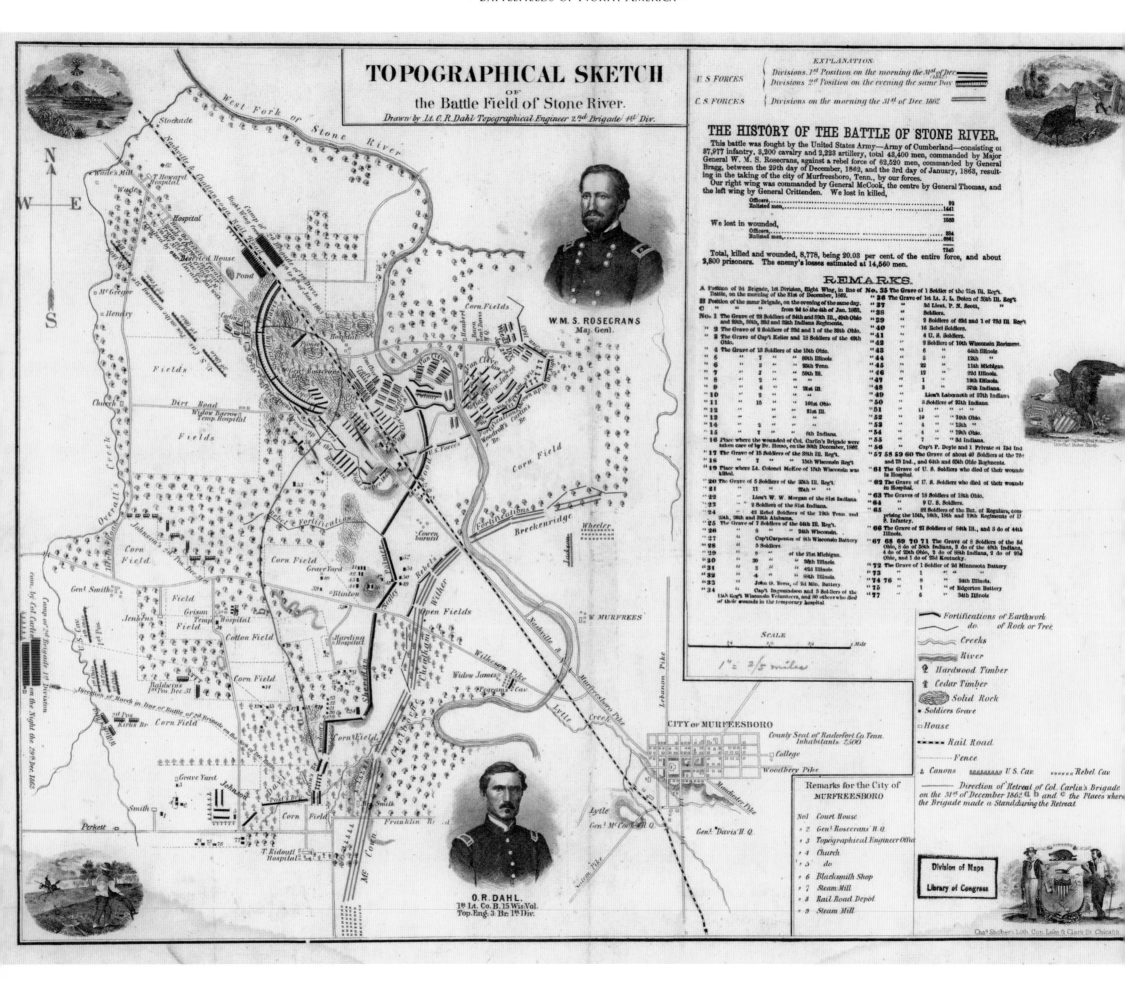

at the end of the first week of November.

For the next several weeks, and much to the chagrin of their respective political masters, neither general seemed in a hurry to strike back at their opponent. Rosecrans was finally prodded into action towards the end of December and advanced the thirty miles from his base to Murfreesboro. Both generals decided to attack and both chose to do so with troops on their respective left wings. Bragg moved first, at around dawn on the 31st, and his advance caught the troops on the Union right flank by surprise.

LEFT: The Confederate forces defending Murfreesboro (their name for the battle), and the important Nashville & Chattanooga Railroad took up positions to the northwest and west of the town astride Stones River. Their Army of Tennessee comprised two corps under Lieutenant-General William Joseph Hardee and Lieutenant-General Leonidas Polk, but a single division under Major-General John McCowen also participated in the action.

ABOVE: Confederate Major-General John Cabell Breckinridge fought in several of the major battles in the western theater, including Shiloh, Vicksburg, Chickamauga, and Missionary Ridge as well as at Stones River.

RIGHT: Major-General Thomas L. Crittenden, whose forces were part of General Rosecrans' army and were heavily engaged in the Battle of Stones River.

RIGHT: Union troops in the foreground advance over a much-contested piece of ground during the first day of the battle. The Confederates largely took the lead during the opening phase of the battle, but their attempt to turn their opponents' right was thwarted after several hours of intense combat.

RIGHT: Irish-born Major-General Patrick Ronayne Cleburne, dubbed the "Stonewall Jackson of the west," originally served in the British Army but emigrated to the United States in 1849. Siding with the Confederacy in 1861, he fought in many of the western theater's major battles, including Shiloh (1862) Stones River (1862–63), Chickamauga, and Chattanooga (both 1863). He was killed in action at Franklin, Tennessee, in November 1864.

FAR RIGHT: A topographical map of the battlefield that shows the Union forces in considerable detail but one that provides only limited information on the Confederate dispositions. This was clearly the inspiration for the colored map on page 89, since the details and information are broadly similar, and both these maps have similarities with that on page 94.

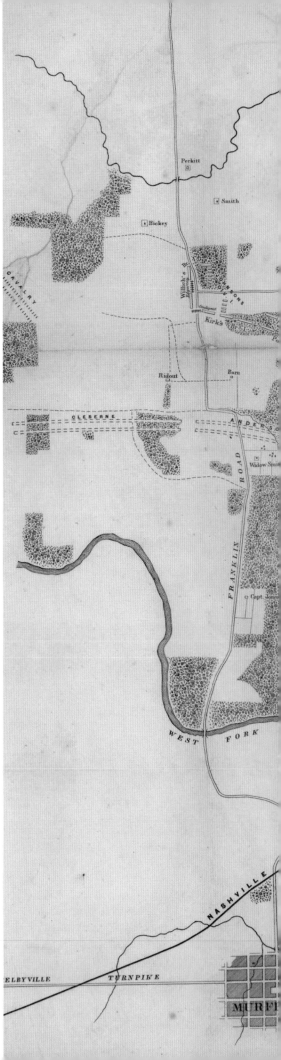

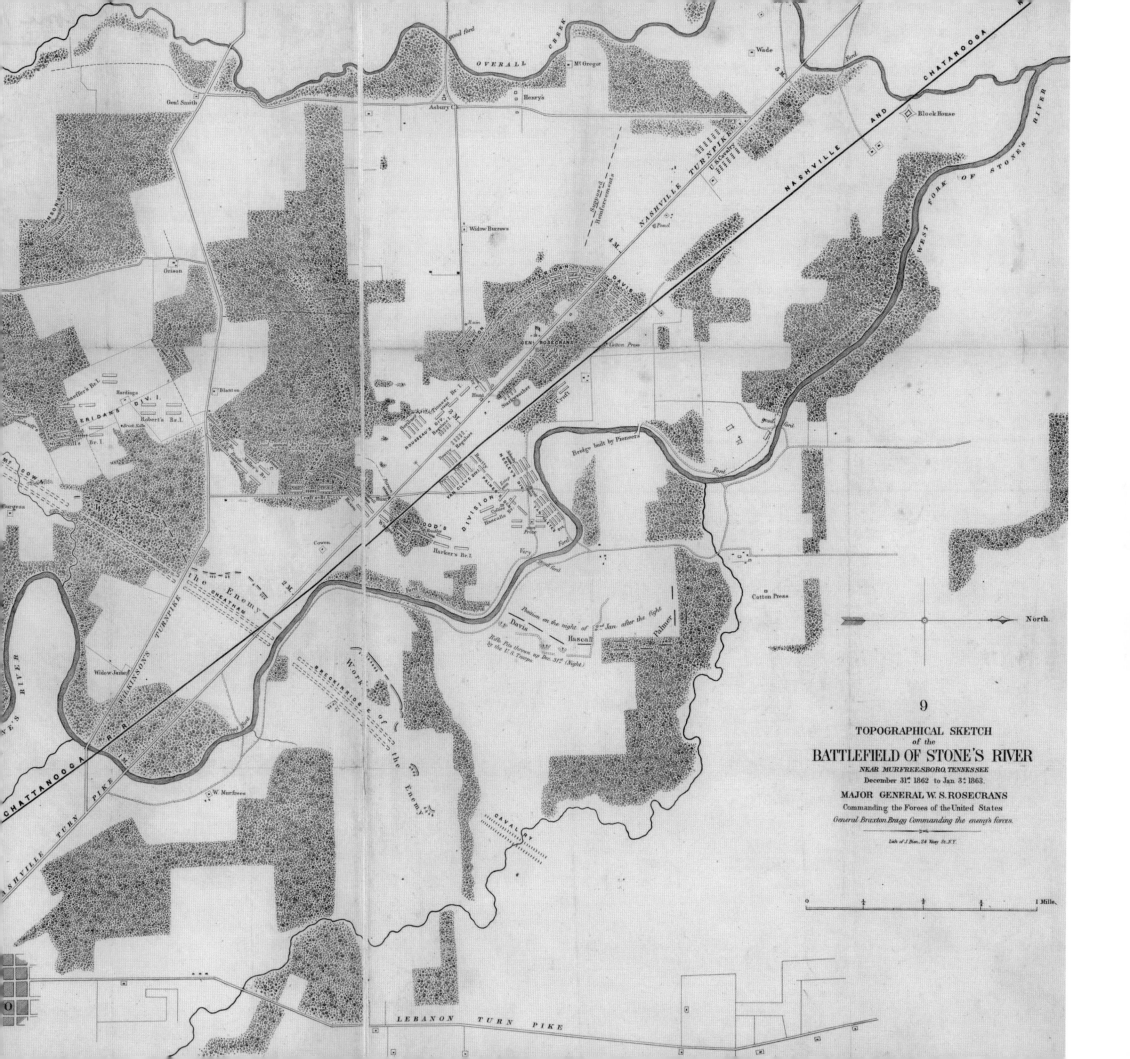

9

TOPOGRAPHICAL SKETCH
of the
BATTLEFIELD OF STONE'S RIVER
NEAR MURFREESBORO, TENNESSEE
December 31st 1862 to Jan 3d 1863.

MAJOR GENERAL W. S. ROSECRANS
Commanding the Forces of the United States
General Braxton Bragg Commanding the enemy's forces.

Lith of J. Bien, 24 Vesey St. N.Y.

North.

0   1 Mile.

OVERALL   CREEK

good ford

Wade

Ford

NASHVILLE AND CHATANOOGA

Mc Gregor

Henry's

Gen'l Smith

Asbury Ch.

Block House

U.S. Cavalry

NASHVILLE TURNPIKE

U.S Cavalry

Widow Burrows

Support of Reinforcements

4 M.

Pond

WEST FORK OF STONES RIVER

Grison

DAVIS

Cotton Press

Ruin

GEN'L ROSECRANS

Blanton

Schaeffer's Br.I.

Hardings

SHERIDANS DIV. I

Robert's Br.I.

Brick Kiln

's Br.I.

Hunt

Stackweather

Croft.

Pioneer Br. I.

ROUSSEAU DIV.I

Bridge built by Pioneers

Ford

Mc COWN

Burgess

WORK

Cruft's Br I.(Right)

VAN CLEVE DIV.

DIVISION

PALMER'S

Hascall Br I.

good ford

WOOD'S

Harker's Br. I.

Cowen.

Very good ford

Ford

Cotton Press

the Enemy

CHEATHAM

2 M.

Position on the night of 2d Jan after the fight

Davis

Hascall

Palmer

NASHVILLE TURNPIKE

CHATTANOOGA R. R.

Rifle Pits thrown up Dec. 31st (Night) by the U.S. Troops

Widow James

Work of the Enemy

BRECKINRIDGE of the Enemy

W. Murfrees

CAVALRY

NASHVILLE TURN PIKE

LEBANON TURN PIKE

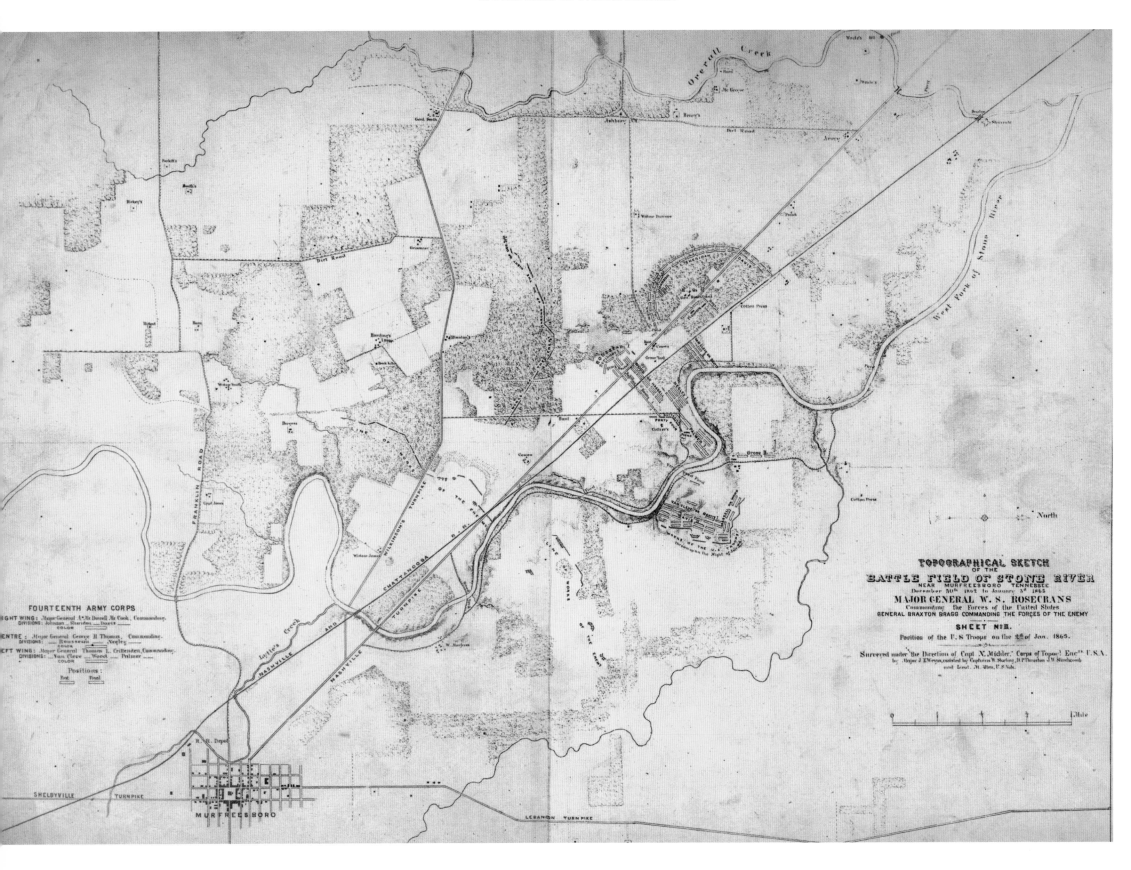

ABOVE: The brunt of the initial Confederate assault was born by three divisions on the Union right under Brigadier-General Alexander McDowell McCook, some 16,000 men in all who faced around 10,000 Southern troops. Two of the Union divisions were caught preparing breakfast and simply crumbled, but the third under Brigadier-General William Tecumseh Sherman was made of sterner stuff and fought back until its ammunition ran low and his opponents threw ever more men into the action.

RIGHT: Union troops hurrying forward to prevent a Confederate breakthrough during fighting along the banks of Stones River. The battle ended when the Federals decided to withdraw. The fighting had been so costly that there was little likelihood of Rosecrans following up his narrow strategic victory. Both sides would refrain from major operations for several months, returning to full-scale action only during the brief Tullahoma campaign from 23 June to 2 July.

It simply fell apart in rapid retreat. Rosecrans' center was made of sterner stuff, especially the division led by Brigadier-General Philip Sheridan. By nightfall the Union commander was able to patch together a new defensive position along the line of the Nashville Turnpike.

The night was bitterly cold and the men were exhausted. It was hardly surprising that the two armies merely held their ground, and no major combat took place on New Year's Day. There was a briefly flurry of action on 2 January, when two divisions clashed on Bragg's extreme right wing. The initial Union advance across Stones River was halted by the arrival of a Confederate division, but it too was repulsed by concentrated fire from some sixty Union field guns as it, in turn, tried to cross the river.

Bragg was informed on the 3rd that Rosecrans was about to receive substantial reinforcements, and he therefore decided to withdraw. The movement began that night but Rosecrans decided against a pursuit. He ordered his troops to dig extensive field fortifications around Murfreesboro, and Bragg did likewise at his new base. Stones River was a tactical and strategic victory for the Union at a time when they were few and far between. Despite the heavy casualties, it gave a boost to the wider public's morale in the North.

THE BATTLE OF CHANCELLORSVILLE, 1863

DATE: 1–6 May 1863

COMMANDERS: (Union) Major-General Joseph Hooker; (Confederate) General Robert E. Lee

TROOP STRENGTHS: (Union) 130,000; (Confederate) 60,000

CASUALTIES: (Union) 16,754 killed, wounded or missing; (Confederate) 12,754 killed, wounded or missing

Major-General Hooker ("Fighting Joe") took command of the Army of the Potomac in January 1863 and, as a commander with a supposed unquenchable thirst for action, decided to launch his command against Lee's Army of Northern Virginia in a complex spring offensive, one designed to turn Lee's left flank at Fredericksburg, Virginia. The advance began on 29 April, when Hooker led some 90,000 men across the Rappahannock River and took them into an area of very dense forest with few roads, ominously known as the Wilderness. He left a further 40,000 men under General John Sedgwick, positioned directly opposite Lee at Fredericksburg.

RIGHT: Union troops are evacuated across the Rappahannock River, probably at Fredericksburg, after the Battle of Chancellorsville had ended on 6 May 1863. The Army of the Potomac's attempt to trap the South's Army of Northern Virginia against the river was thwarted and it was the Federal force that came closest to disaster, largely due to a rapid flank march by troops under Lieutenant-General Thomas "Stonewall" Jackson.

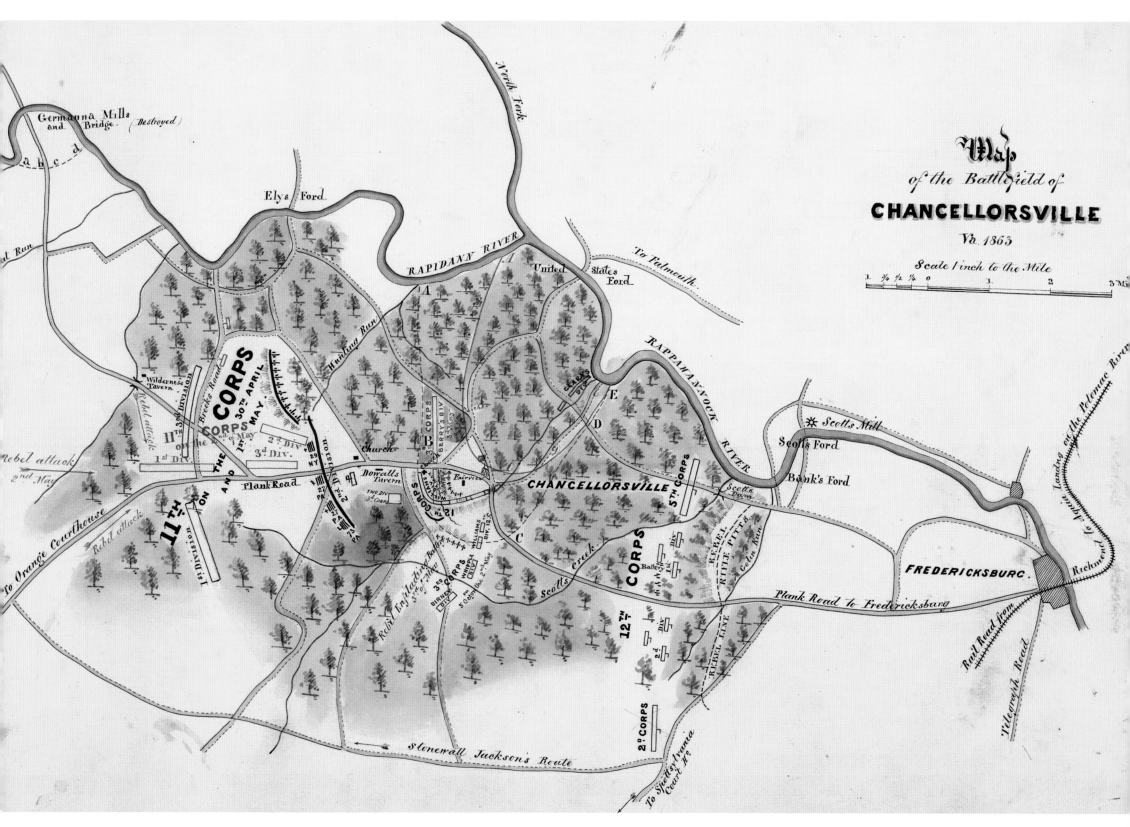

Map
of the Battlefield of
CHANCELLORSVILLE
Va. 1863

Scale 1 inch to the Mile

ABOVE: The Union commander, Major-General Joseph Hooker, had sent the bulk of his Army of the Potomac swinging around to the rear of the Confederate forces and hoped to bring them to battle in the mostly open ground to the west of Chancellorsville so that his greater numbers of artillery batteries would tell. However, Hooker somewhat panicked when his lead troops ran into a line of Confederate skirmishers, and his troops were ordered to pull back into an area of thick forest with tangled undergrowth known as the Wilderness. It was not an auspicious start and even Hooker subsequently lamented that, "For once I lost confidence in Hooker."

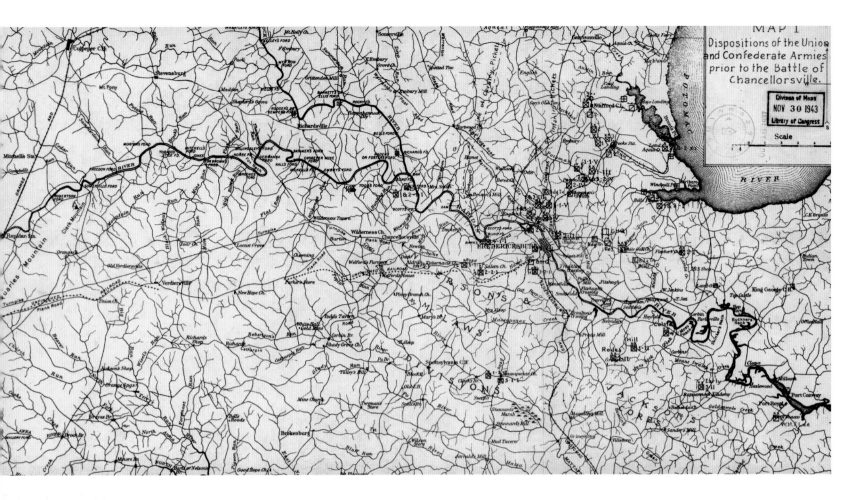

LEFT: The location of the various Union and Confederate forces around Fredericksburg before the beginning of the battle on 1 May. The bulk of the Army of the Potomac, some 80,000 men split between five corps, struck camp on 24 September and began a march around the left flank of the Confederates. After fording the Rappahannock and Rapidan Rivers at various points, the Union troops were largely concentrated at Chancellorsville by the 30th.

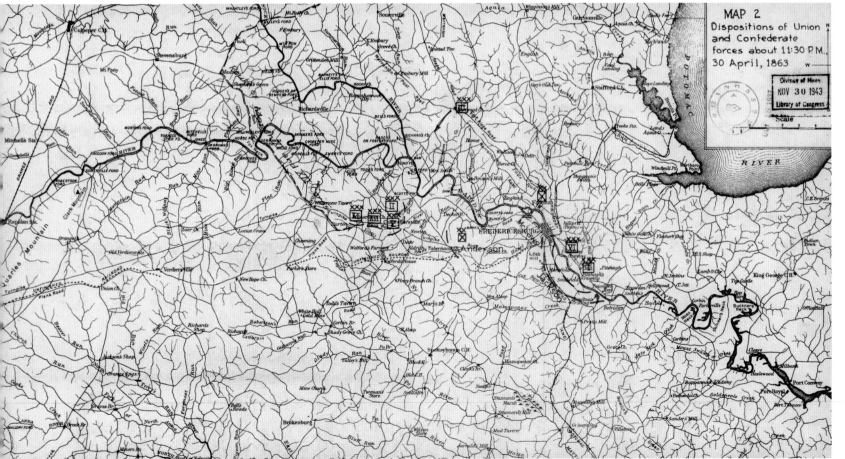

LEFT: The disposition of various Union and Confederate forces on 30 April. As the Union forces pushed eastward from Chancellorsville, General Robert E. Lee split his Army of Northern Virginia, leaving a holding force of some 10,000 men to screen Fredericksburg while the next day moving another 50,000 men towards the oncoming Union troops to his rear.

RIGHT: Major-General James Ewell Brown "Jeb" Stuart performed several key functions during the battle—on 30 April he informed Lee that the Union forces around Chancellorsville were marching on his rear, and on 1 May reported to him that the Union right flank was relatively exposed.

FAR RIGHT: A studio portrait of several Union officers. At the extreme left is Howard, who commanded the ill-fated XI Corps at Chancellorsville and Gettysburg. Also present is Brigadier-General William Tecumseh Sherman (center, seated). The image was most likely taken between May–December 1864 either during the Atlanta campaign or during the subsequent "March to the Sea" through Georgia, when Howard was serving under Sherman.

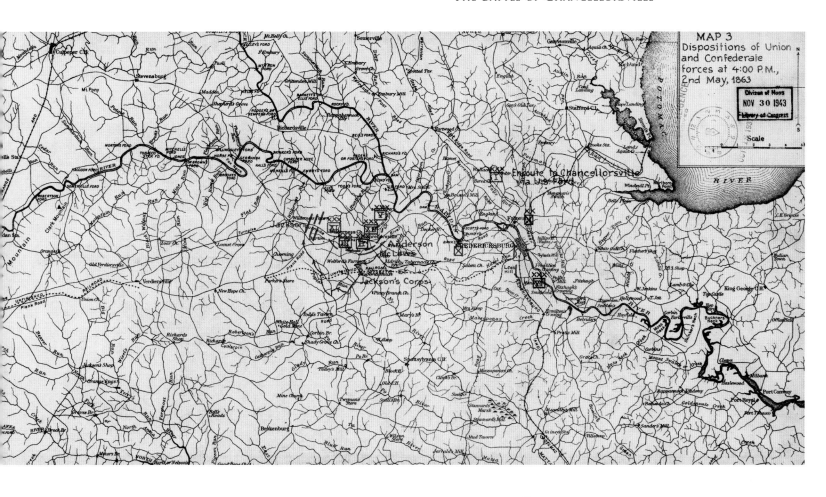

MAP 3
Dispositions of Union
and Confederate
forces at 4:00 P.M.,
2nd May, 1863

LEFT: The Battle of Chancellorsville on the afternoon of 2 May. The Union push on Fredericksburg had ground to a halt the previous day, for little good reason, allowing Lee and Jackson to plot the latter's daring march against the exposed right flank of the Union forces. The blow came at around 1715 hours when about 26,000 Confederate troops hit Major-General Oliver Otis Howard's XI Corps, a formation with a high proportion of recent German immigrants who did not speak English. The corps, which contained around 9,000 men, simply melted away in panic.

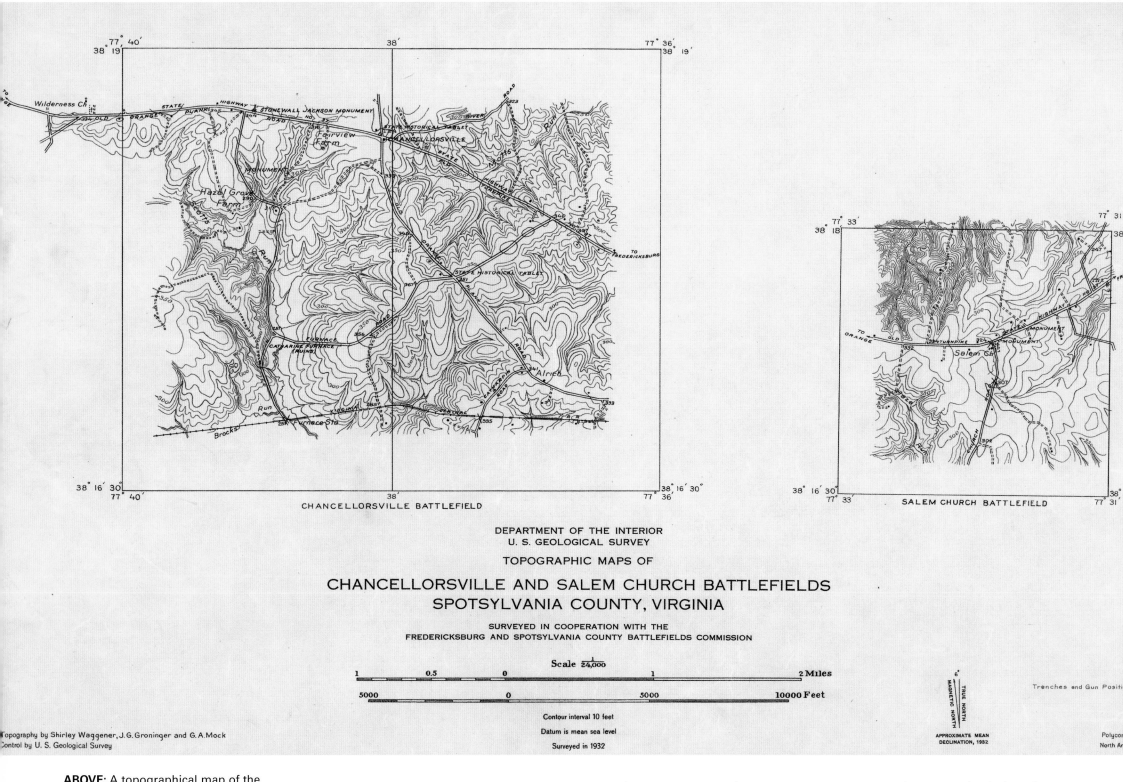

CHANCELLORSVILLE BATTLEFIELD

SALEM CHURCH BATTLEFIELD

DEPARTMENT OF THE INTERIOR
U. S. GEOLOGICAL SURVEY

TOPOGRAPHIC MAPS OF

CHANCELLORSVILLE AND SALEM CHURCH BATTLEFIELDS
SPOTSYLVANIA COUNTY, VIRGINIA

SURVEYED IN COOPERATION WITH THE
FREDERICKSBURG AND SPOTSYLVANIA COUNTY BATTLEFIELDS COMMISSION

Scale 1/24,000

Contour interval 10 feet

Datum is mean sea level

Surveyed in 1932

Topography by Shirley Waggener, J.G. Groninger and G.A.Mock
Control by U. S. Geological Survey

Trenches and Gun Position

APPROXIMATE MEAN
DECLINATION, 1932

Polyconic

North Am

ABOVE: A topographical map of the southern portion of the battlefield of Chancellorsville. The spot where Jackson was accidentally shot by his own troops is marked by a monument (top left) to the west of Chancellorsville. His key flank march took him through Catharine Furnace (bottom left) and then farther south and west before he turned to the north.

Lee's response was to leave some 10,000 men under General Jubal Early to cover Sedgwick and make a staged withdrawal if the latter decided to attack, then ordered the remainder of his command to move against Hooker's main force. Lieutenant-General Thomas "Stonewall" Jackson was told to circle around Hooker's right flank outside Chancellorsville, by way of Wilderness Tavern, while Lee with some 17,000 troops would take a more direct route towards Hooker's center.

The first contact came on 1 May and, despite his superior numbers, Hooker quickly lost his nerve and went onto the defensive. Jackson began his fifteen-mile

LEFT: Massachusetts-born Major-General Joseph "Fighting Joe" Hooker was roundly defeat at Chancellorsville and quickly tended his resignation as commander of the Army of the Potomac. He subsequently fought as a corps commander at Chattanooga and played a pivotal role in the Battle of Lookout Mountain on 24 November 1863. While an undoubtedly able administrator, he was not psychologically suited to the highest command, but he was nevertheless a fine subordinate commander.

BELOW: Stuart poses for a formal portrait. During the battle, he took charge of the mortally wounded Jackson's command on 3 May and immediately launched it against the Union's III Corps under Major-General Daniel E. Sickles. The latter was caught as he moved back to friendly lines, and Stuart was able to push the III Corps off the high ground at Hazel Grove Farm (see map opposite top left), from where thirty Confederate cannon were subsequently able to pound Chancellorsville.

march at 0800 hours the next day and launched a ferocious assault on Hooker's right flank, as planned, at a little before dusk. The Union troops were overwhelmed and the whole wing just disintegrated. Jackson sensed that total victory might be in sight and went forward to assess the situation, but was fired on and mortally wounded by his own troops. His command went to Major-General J. E. B Stuart.

The 3rd saw Sedgwick push Early out of Fredericksburg and then try to move on Chancellorsville. Lee correctly sensed that Hooker was so knocked of balance by the recent events that he would largely remain inactive, so led some 25,000 men to aid Early. Sedgwick came close to being cut off near Salem Church but just managed to cross back over the Rappahannock on the 4th. Hooker followed him the next day and completed the retreat on the 6th. Chancellorsville was undoubtedly Lee's masterpiece,

ABOVE: Men from Major-General John Sedgwick's VI Corps await orders in entrenchments below Marye's Heights at Fredericksburg, on the west bank of the Rappahannock River. Sedgwick commanded 40,000 men and was tasked with fixing the Army of Northern Virginia at Fredericksburg while Hooker made his flank march in late September. However, Lee soon withdrew the bulk of his forces from around Fredericksburg, leaving just 10,000 men under General Jubal Early to oppose Sedgwick.

one in which he comprehensive outmaneuvered his more timid opponents so as to offset his significantly lower number of troops.

Nevertheless, the battle had cost the Army of Northern Virginia dear and the Confederacy did not have bottomless reserves of manpower. Jackson, who died of his wounds on the 10th, was irreplaceable. Hooker proposed a second two-pronged attack on Lee in June but his chastened superiors feared a second Chancellorsville, and vetoed his plan. The general tendered his resignation and it was accepted on the 28th, just days before the crucial Battle of Gettysburg.

LEFT: Connecticut-born Major-General John Sedgwick was much liked by his men, who nicknamed him "Uncle John," but he was a fighting general. In 1862, for example, he was wounded during the Peninsular Campaign and twice wounded during the Battle of Antietam. He commanded the Union left wing on the third day at Gettysburg the following year but was killed by a Confederate sharpshooter at the Battle of Spotsylvania, fought during 8–18 May 1864. He was on an inspection tour of Union entrenchments that were under fire from a distance of up to 800 yards by Confederate marksmen equipped with imported British Whitworth rifles fitted with telescope sights. The general ignored repeated warnings and refused to duck, even repeatedly quipping that, "They couldn't hit a barn door at this distance." Moments later he was hit just below the left eye and quickly bled to death.

THE BATTLE OF GETTYSBURG, 1863

DATE: 1–3 July 1863

COMMANDERS: (Union) Major-General George Meade; (Confederate) General
Robert E. Lee

TROOP STRENGTHS: (Union) 88,000; (Confederate) 75,000

CASUALTIES: (Union) 3,115 killed, 14,529 wounded, and 5,365 missing;
(Confederate) 2,500+ killed, c.13,000 wounded, and 5,425 missing.

The Battle of Chancellorsville in May 1863 had given Lee the strategic initiative and he determined to launch yet another invasion of the North. His Army of Northern Virginia crossed the Potomac River into the North between 15 and 24 June. The commander of the Union's Army of the Potomac, Major-General Joseph Hooker, began to move his forces north to intercept Lee but repeated missed opportunities to hit the Confederate leader while his various corps were strung out on the line of march. When Hooker's plan of attack was opposed, he resigned on the 28th to be

RIGHT: The first day of Gettysburg, 1 July, effectively began as a battle of encounter to the immediate north of the town, in which elements of both sides rather blundered into each other at around 0800 hours. Reinforcements then gradually arrived over the following hours. This map shows the relative positions of the Union Major-General John Fulton Reynolds' I Corps, which arrived at around 1000 hours, and Major-General Oliver Otis Howard's XI Corps. Both were originally opposed only by the Confederate III Corps under Lieutenant-General Ambrose Powell Hill but were forced to retreat to the south of Gettysburg due to the arrival of Lieutenant-General Richard Stoddert Ewell's II Corps on their exposed right flank. Reynolds was killed shortly after arriving and was temporarily replaced by Major-General Abner Doubleday, an officer who had fired the first Union cannon during the bombardment of Fort Sumter back in 1861.

LEFT: The area over which the Gettysburg campaign was fought. Lee sent his three corps away from Fredericksburg at the end of May, hastening them on separate looping drives northwards that took them on a march from Virginia into West Virginia's Shenandoah Valley, and from there into Pennsylvania by way of the Cumberland Valley. At various points Lee's corps then began to turn south with the intention of menacing Washington from the north.

FAR LEFT: Confederate troops of General James Longstreet's I Corps are depicted during the so-called "Pickett's Charge" against the center of the Union line on Cemetery Ridge on the hot afternoon of 3 July. The fiercest hand-to-hand fighting developed at a point christened the "Bloody Angle," a point roughly where the clump of tress is shown here. Over half of those who made the charge—lasting just sixteen minutes or so—were killed, wounded, or captured.

RIGHT: A computer-generated topographical map of the battlefield. Big Round Top and Little Round Top are unmistakeable on the Union left flank, as is Cemetery Ridge, which runs northwards from the two promontories.

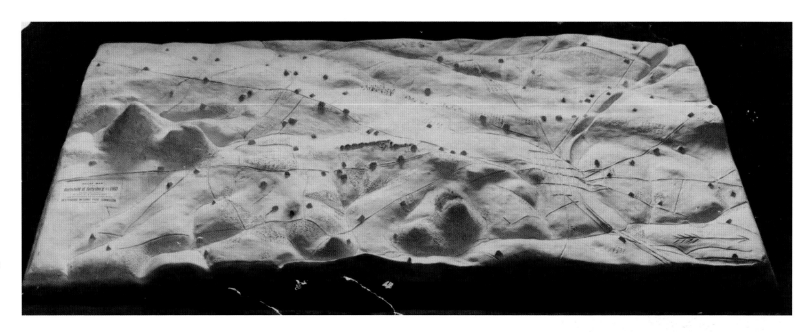

BELOW: A panoramic view of the battlefield that clearly shows how the Union line formed something of a fish-hook shape, with Culp's Hill rising above Rock Creek (below left) forming the barb. It then swings a little way west before running southward along Cemetery Ridge before ending at the twin Round Tops.

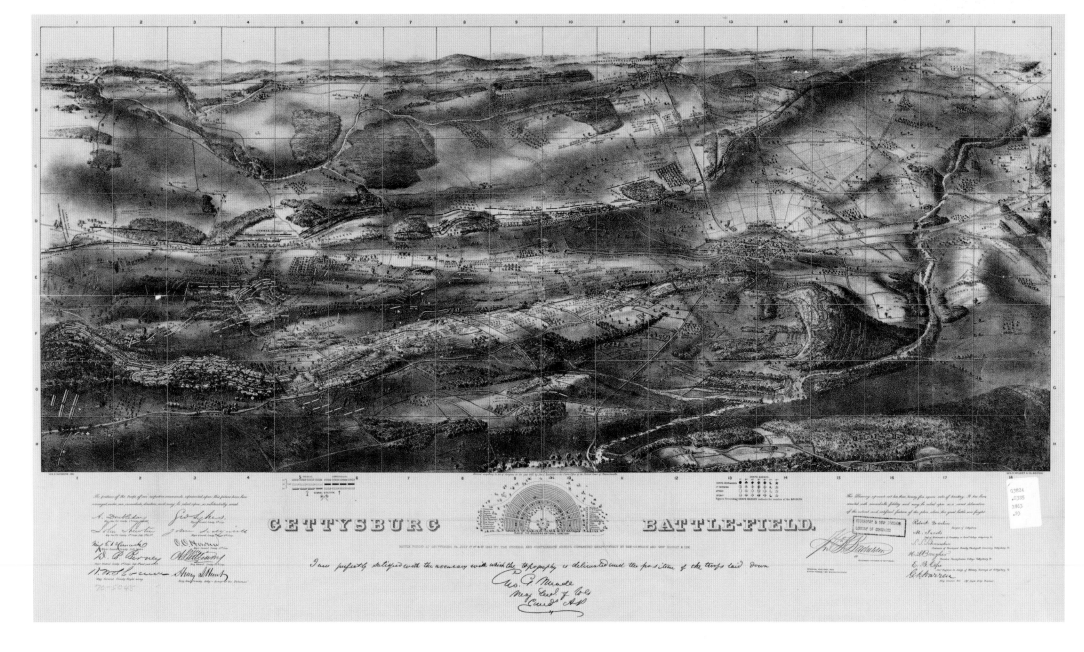

ABOVE: Major-General Winfield Scott Hancock commanded the Army of the Potomac's II Corps at Gettysburg and was tasked with defending Cemetery Ridge during "Pickett's Charge" on the third day of the battle.

RIGHT: This map shows the situation at Gettysburg on the various days of the battle. The fighting effectively spread southward from the town as the arriving Union forces were positioned on the high ground of Cemetery Ridge.

BELOW: Major-General John Fulton Reynolds commanded the Union's I Corps, the first of the Army of the Potomac's infantry corps to arrive at Gettysburg on 1 July.

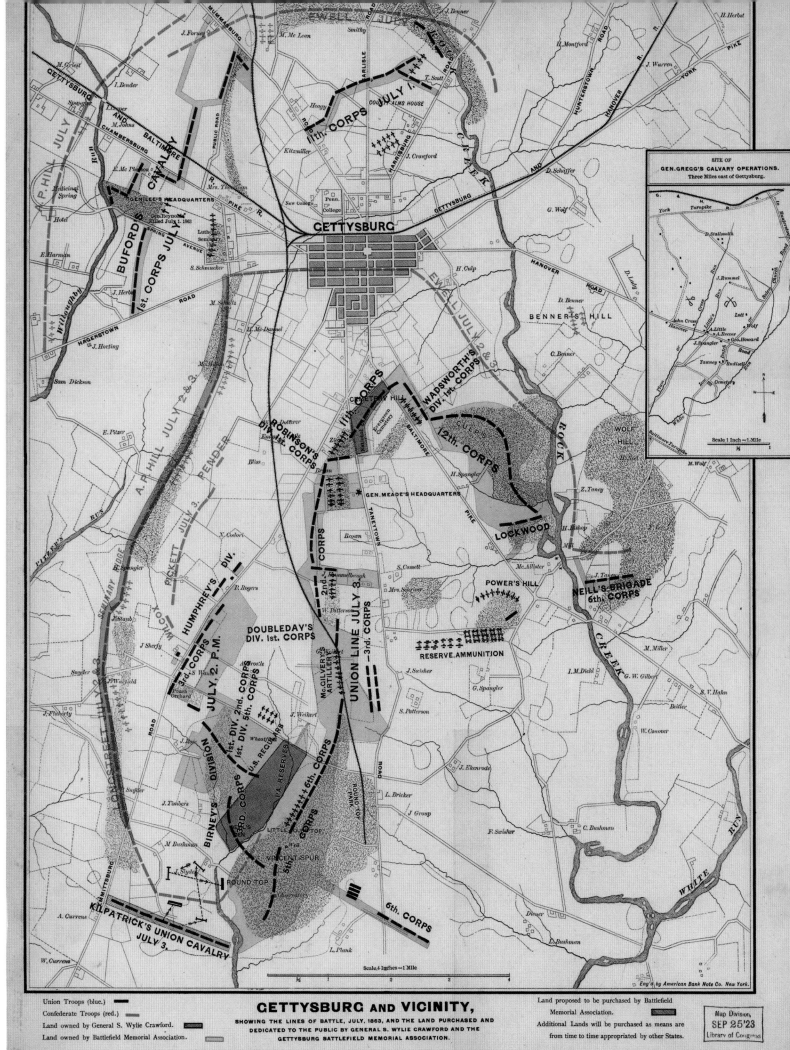

BELOW: A rough map of Gettysburg area shows relative positions of both Union and Confederate armies during the second and third days of the action. An inset map shows the layout of burials in the National Cemetery, where President Abraham Lincoln gave his 272-word "Gettysburg Address" on 19 November 1863.

RIGHT: This map of the battle emphasizes the role of artillery during the three days of fighting. It also shows two of the lesser-known events of the last day. In the southwest, Union cavalry tried to turn the Confederate right flank, while the latter attempted to do the same to the Union right in the northeast.

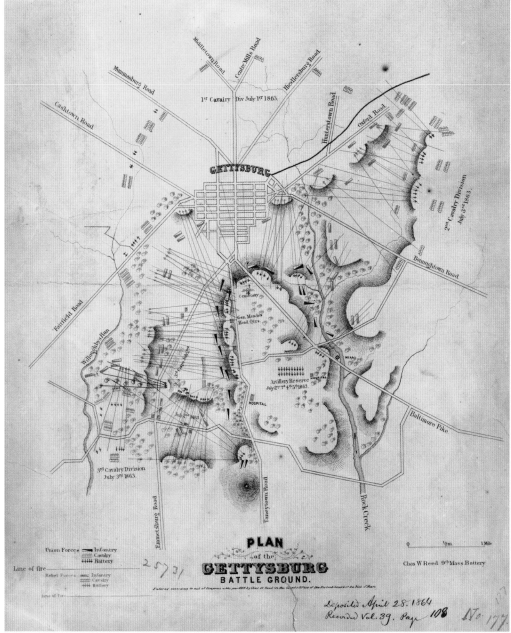

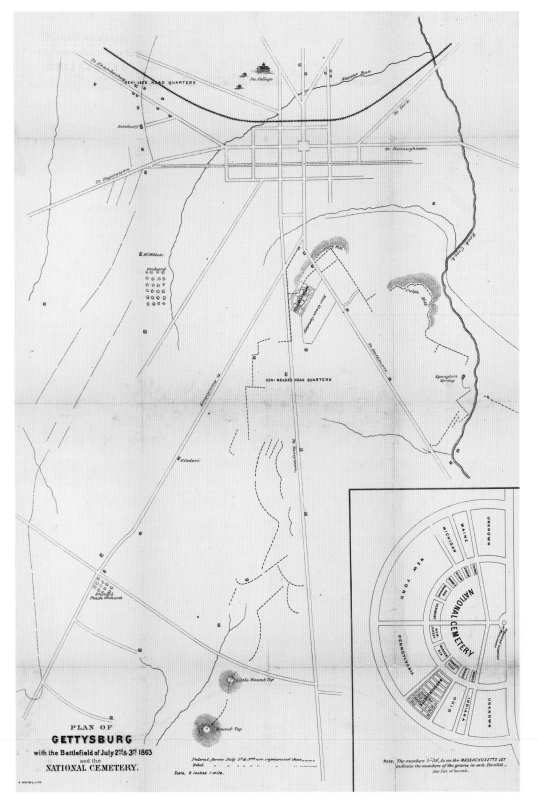

replaced by Meade—the army's fifth commander in just ten months.

Lee was supposedly meant to be receiving intelligence from Major-General J. E. B. Stuart's cavalry, but Stuart had embarked on a raiding mission on the 26th and would be missing until 2 July. Nevertheless, there were some vague reports of the Army of the Potomac's whereabouts. A Union cavalry brigade undertaking a reconnaissance was moving cautiously northwards in Maryland and happened to bump into a Confederate infantry brigade just to the north of Gettysburg on the 30th. Both sides now tried to concentrate their forces there.

The battle proper opened on 1 July to the north of the town, with both sides feeding troops into the action as they arrived. Lee had wanted to make a powerful offensive push into the Union then secure good defensive ground for his entire force and wait for his opponents to attack. By the end of the first day, it was the Union troops who were holding good defensive ground to the south of Gettysburg. Lee knew that the onus would be on him to attack, but his army was still arriving.

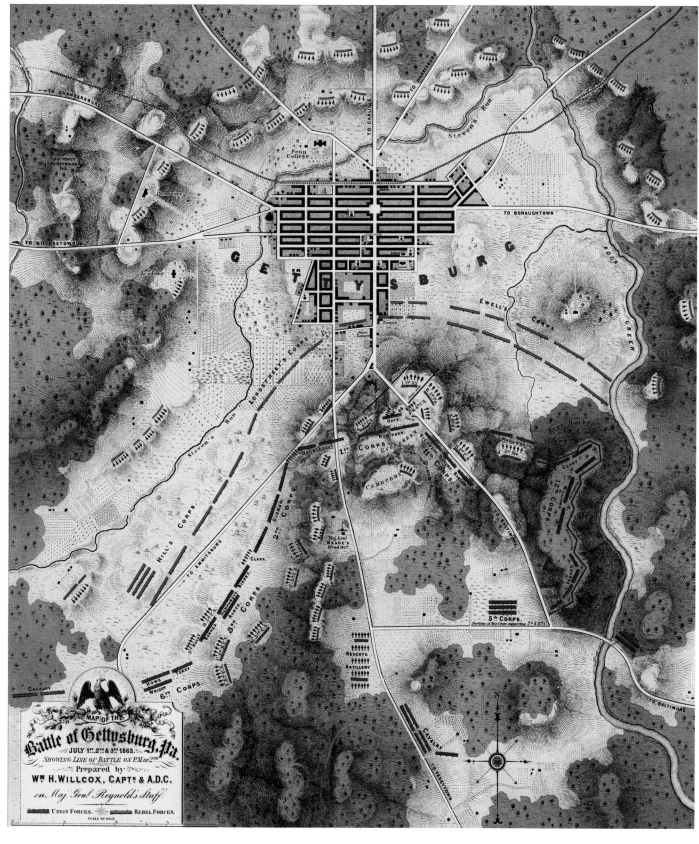

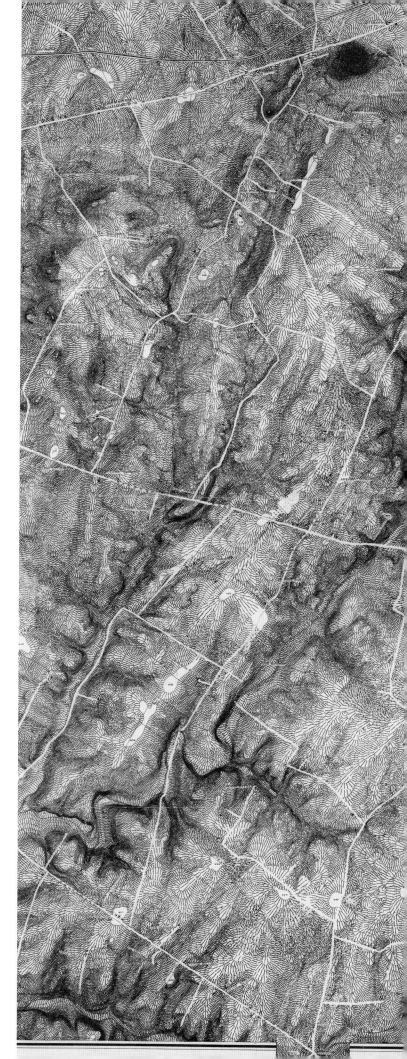

ABOVE: The rival dispositions during the second day of the battle, with three corps arcing around the Union positions from the northeast to the southwest.

RIGHT: The area in the northeast of the battlefield where Union and Confederate cavalry clashed on the final day of the fighting. Although many of the troopers fought dismounted, there was a major charge involving Union Brigadier-General George Armstrong Custer's 3rd Cavalry Brigade and the Confederate 1st and 4th Cavalry Brigades under Brigadier-Generals Wade Hampton and Fitzhugh Lee.

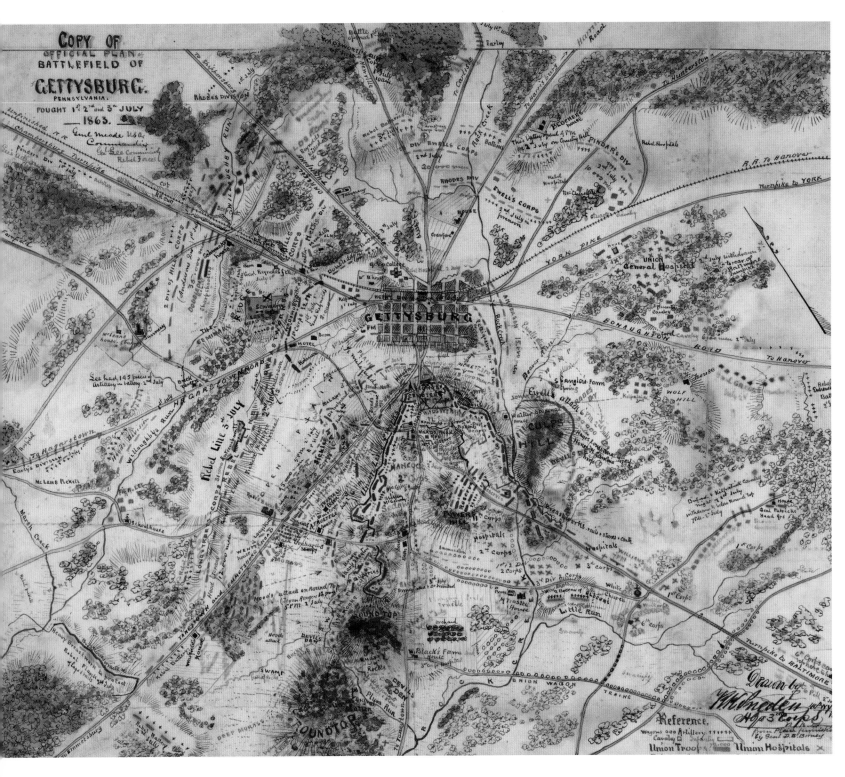

LEFT: A somewhat confusing map of Gettysburg that attempts to cover the course of the fighting over three days. It also shows the locations of the Army of the Potomac field hospitals, which are shown concentrated to the southeast of the main battlefield and close to the main turnpike to Baltimore.

RIGHT: A topographical map of the area surrounding Gettysburg, one that shows that the Maryland town was the focal point of several roadways. The Confederate forces moved towards Gettysburg from the northeast, chiefly along the Chambersburg Pike, Cashtown Road, and Mummasburg Road.

On day two Lee decided that he would try to swing round the far left of the fish-hook-shaped Union line and also strike against its extreme right flank. The attack against the left began in the afternoon once the corps needed had arrived. The Union troops holding positions along the Emmitsburg Road were forced back to the summit of Cemetery Ridge, but the attackers were thwarted in turning the far left flank by the staunch defense of two high points, Little and Big Round Top. The move against the right made some progress, but the gains were abandoned when no reinforcements arrived.

For day three Lee, who was poorly, made a fateful decision to punch his way through the very center of the Union line—an uncharacteristically unimaginative strategy that some of his generals opposed—and some 12,000–15,000 troops were committed to what became known as "Pickett's Charge." Many Rebel troops failed to reach the Union positions on Cemetery Ridge, being cut down by intense rifle and cannon fire. Even fewer actually crossed the low stone wall that marked the front line. Within minutes the shattered Confederate units broke and fled, but no more than 5000 returned to their own lines. The battle was effectively over—the "high tide of the Confederacy" had briefly arrived, but rapidly departed on Cemetery Ridge. It was only a matter of time before the South was defeated.

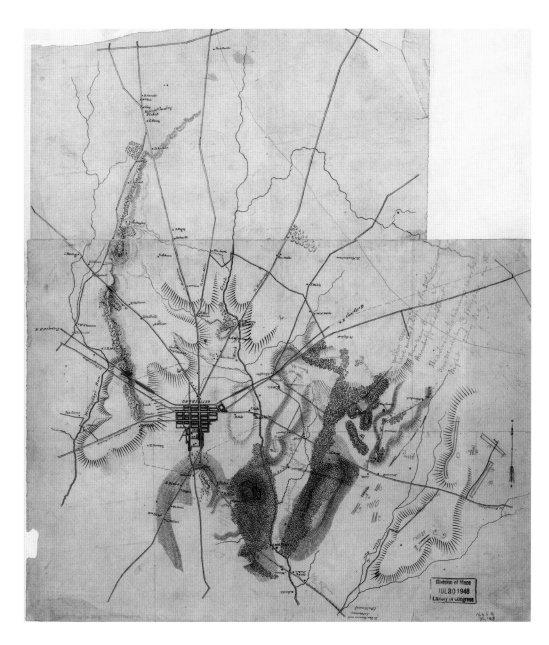

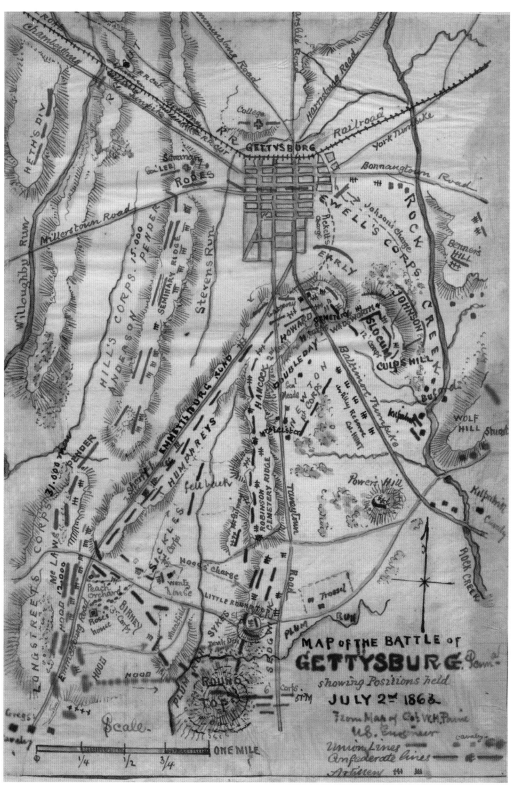

LEFT: Just a handful of the battle's casualties—these appear to be Union dead. There were so many corpses that most were buried in communal graves.

ABOVE: The disposition of various units on the second day of the battle, 2 July. The fighting focused on the Union left, around Little Big Top, which was unsuccessfully assaulted by Major-General Lafayette McLaws' 1st Division and Major-General John Bell Hood's 3rd Division.

THE BATTLE OF CHICKAMAUGA, 1863

DATE: 19–20 September 1863

COMMANDERS: (Union) Major-General William S. Rosecrans; (Confederate) General Braxton Bragg

TROOP STRENGTHS: (Union) 62,000; (Confederate) 65,000

CASUALTIES: (Union) 1,656 killed, 9,756 wounded, and 4,757 missing; (Confederate) 2,312 killed, 14,674 wounded, and 1,486 missing

Aside from Grant's ongoing struggle to capture Vicksburg, there was little action elsewhere in the civil war's western theater during the first six months or so of 1863. Rosecrans' Army of the Cumberland remained at Murfreesboro, while Bragg remained stationed at Tullahoma. After his superiors demanded action, Rosecrans made a successful attempt to push Bragg back to Chattanooga between 23 June and 2 July. After another lengthy delay, the Army of the Cumberland moved on Chattanooga aided by the Army of the Ohio, which pushed towards Knoxville, Tennessee, from its base at Lexington, Kentucky.

Bragg was forced to abandon Chattanooga on 8 September and withdrew to

RIGHT: A map of the Chickamauga battlefield and its surroundings. The actual battle took place along Chickamauga Creek (bottom left) with the Confederates attacking across it from right to left. The distinctive turnpike that runs from top to bottom consists of the Rossville and Lafayette Roads.

LEFT: An attempt to capture the drama of the fighting during the Battle of Chickamauga, although much of the two days of action actually took place in densely wooded terrain with no more than small open spaces around various scattered farms. The fight was initiated by the Confederates, who were trying to turn the Union left flank so as to cut it off from its base at Chattanooga (a few miles to the northwest), which had been abandoned by the South on 7 September.

LEFT: Alexander M. McCook had already been routed twice (at Perryville and Stones River), and when this occurred again at the Battle of Chickamauga he was court-martialed, but not convicted, though he saw no further active service at the front for the remainder of the war.

ABOVE: Ohio-born Major-General William Starke Rosecrans, "Old Rosey" to his men, commanded three front-line and one reserve corps at Chickamauga, and his Army of the Cumberland largely comprised regiments drawn from the mid- and northwest states. Despite having scored a precious victory at Stones River (1862–63), defeat at Chickamauga and the retreat back to Chattanooga after the battle would eventually lead to his dismissal.

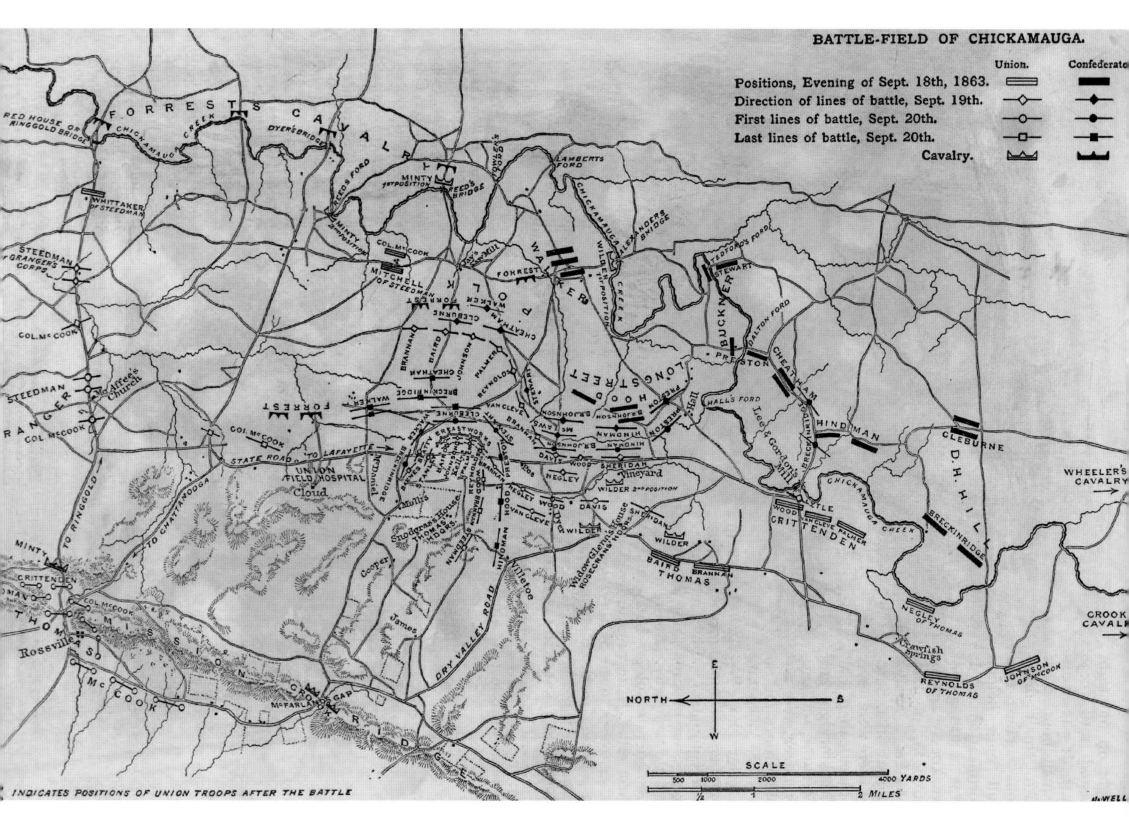

ABOVE: A map showing the terrain and dispositions of the various Confederate and Union corps and divisions over the two days of the fighting, as seen from behind the Federal positions. The crisis point for the Union's Army of the Cumberland came on the second day, 20 November, when units were withdrawn from the center, leaving a gap through which the corps detached from the Army of Northern Virginia, commanded by Major-General James Longstreet, poured around midday.

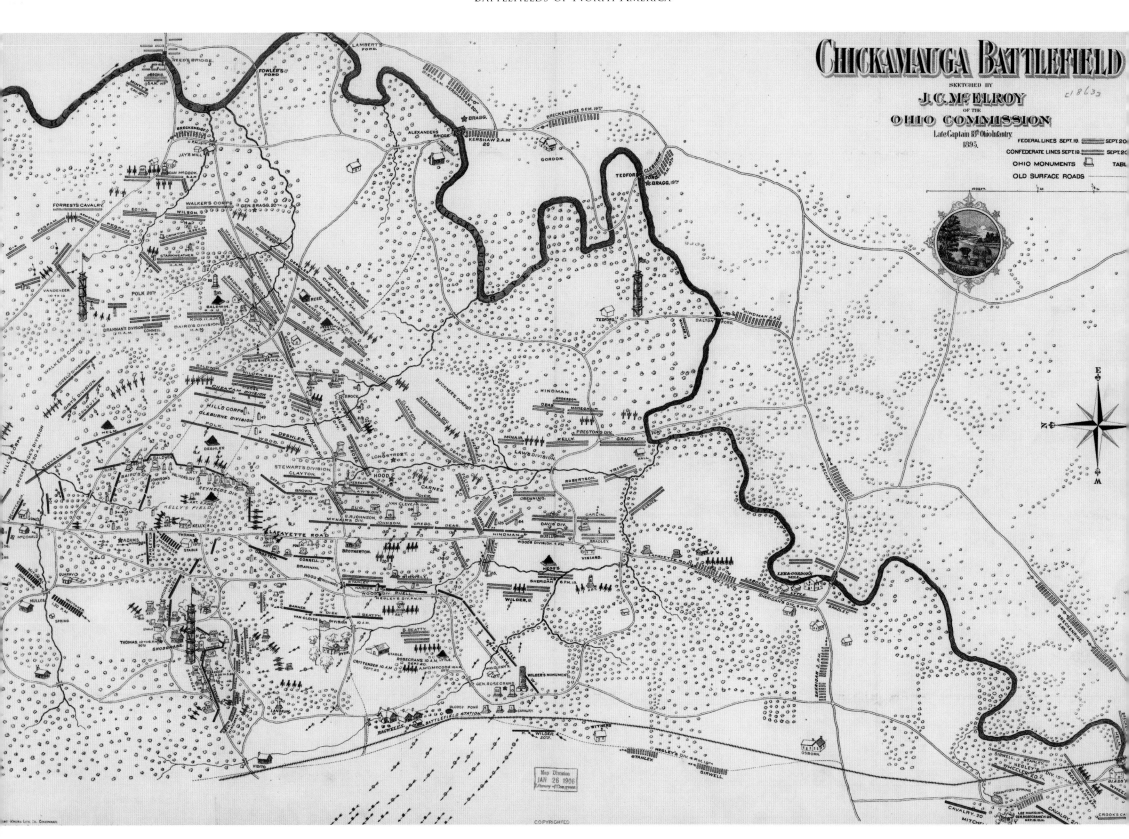

ABOVE: A highly detailed map of the battlefield that takes pains to show just how densely wooded it actually was. General Braxton Bragg's Army of Tennessee is here shown after the greater part of its force had crossed over Chickamauga Creek. Despite having the edge for much of the two-day battle, by the night of the 20th Bragg believed his army had been badly mauled. However, his mood was much improved around midnight when he was visited by Lieutenant-General Leonidas Polk, who informed him that the Army of the Cumberland was "fleeing precipitately from the field." This was something of an exaggeration but the Federals were certainly in retreat back towards Chattanooga.

RIGHT: Benjamin Franklin Cheatham served as major-general of state militia before the Civil War, making it natural that he would be commissioned a brigadier in Confederate service as soon as his home state, Tennessee, seceded. He commanded a division under Bragg during the Battle of Chickamauga and, despite previous reports that he had been seen drunk and not in control of his forces, he was elevated to corps command on 29 September 1863. He was on the right flank of Missionary Ridge when Bragg was defeated by Grant at Chattanooga.

BELOW: James Longstreet was sent by Lee to assist the slothful Bragg, but he did much more than that at Chickamauga, where his habit of coming onto the field of battle at the eleventh hour virtually turned the tide of battle. When Thomas J. Wood's command pulled out of the center of the Federal line, in obedience to confused orders, this opened up a massive hole just as Longstreet'sw men were going in to assault. The whole Union right collapsed and fled for Cahattanooga.

BELOW: Major-General George Henry Thomas, the "Rock of Chickamauga" who largely prevented the total collapse of the Army of the Cumberland during the battle. He was officially the commander of its large four divisions, a 23,000-man XIV Corps but, in fact, operated as the leader of the army's entire right wing.

RIGHT: A Currier & Ives lithograph of the Battle of Chickamauga, showing Thomas (bottom left, mounted) standing firm and counterattacking during the second day of the fighting. Once again, the soldiers are dressed far too neatly.

LaFayette in northwest Georgia. Coming on top of the defeat at Gettysburg on 3 July and the loss of Vicksburg on the next day, the news that Rosecrans was pushing into the heart of the Confederacy along various routes warranted immediate action. A corps from the Army of Northern Virginia was rushed to Bragg's aid and the latter decided to attack Rosecrans, who was still trying to concentrate his various corps outside Chattanooga. Partly due to the densely wooded terrain the two armies eventually stumbled into each other along Chickamauga Creek late on 18 September.

The first day of the battle saw fierce if confused fighting in which neither side gained any clear advantages, but matters changed on the 20th. Rosecrans issued faulty orders that moved some divisions unnecessarily, thereby leaving a gap in his center through which poured the corps of troops from the Army of Northern Virginia. The Army of the Cumberland's center and right flank collapsed and many troops, roughly a third of the total and including Rosecrans, simply abandoned the field. Only the staunch resistance of the surviving left flank under Major-General George Thomas, soon to be dubbed the "Rock of Chickamauga," prevented total disaster.

The "River of Death," so called by Native Amricans, had lived up to its name. Both sides had suffered around 28 percent losses in the two-day struggle around Chickamauga, the biggest battle in the western theater. Despite having the possible opportunity of destroying the greater part of Rosecrans' shattered command on the 20th, Bragg made little immediate effort to pursue. The Union forces fell back to Chattanooga, and Confederate forces placed the city under siege a few days after the battle. The Union forces would remain bottled up until relieved in late October. Rosecrans' career was ruined by that one mistake at Chickamauga; he was relieved of his command to be replaced by Thomas.

RIGHT: Thomas (second from left) gives orders to an aide-de-camp while surrounded by the rest of his staff during the crisis point of the fighting. If his men had not stood fast then the line of retreat back to the Army of the Cumberland's base at Chattanooga would have been severed.

BELOW: A map showing part of the battlefield around Snodgrass Farm and Dyer Field. It was a little to the southeast of here on the second day that the Confederates were able to exploit a gap in the Army of the Cumberland's center and then swing in this direction to reach Horseshoe Ridge which lies just off the south edge of the map. It was there that their attack largely stalled

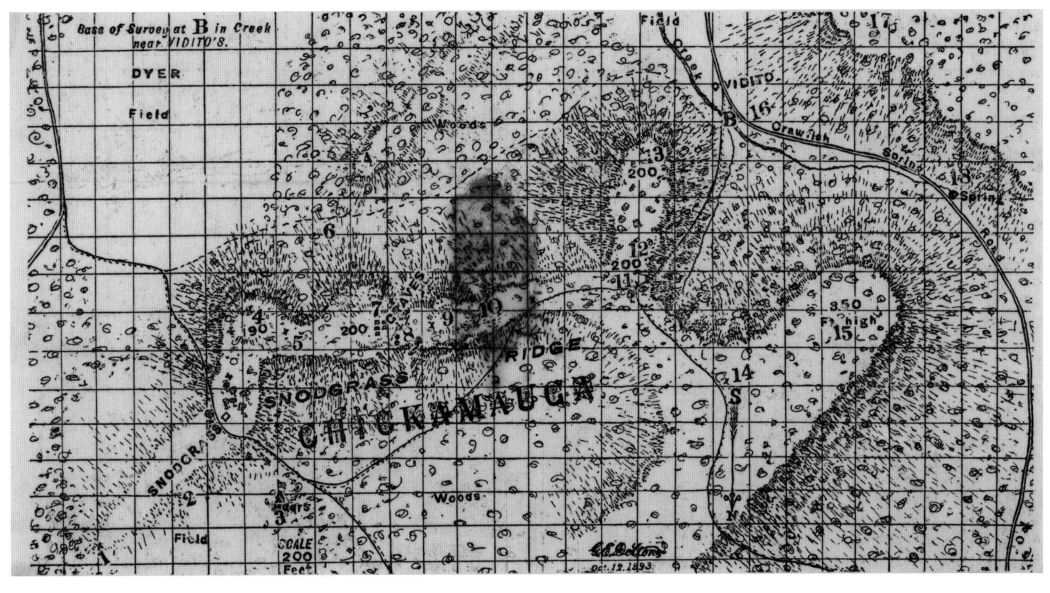

THE BATTLE OF CHATTANOOGA, 1863

DATE: 24–25 November 1863

COMMANDERS: (Union) Major-General Ulysses S. Grant; (Confederate)
General Braxton Bragg

TROOP STRENGTHS: (Union) 61,000; (Confederate) 40,000

CASUALTIES: (Union) 753 killed, 4,722 wounded, and 349 missing;
(Confederate) 361 killed, 2,160 wounded, and 4,146 missing

Following his defeat at the Battle of Chickamauga in September 1863 the commander of the Union's Army of the Cumberland, Major-General William Rosecrans, withdrew into Chattanooga, where he was besieged by Bragg's Army of Tennessee. The situation was stalemated over the next several weeks, despite two corps of the Union's Army of the Potomac, under Major-General Joseph Hooker, being

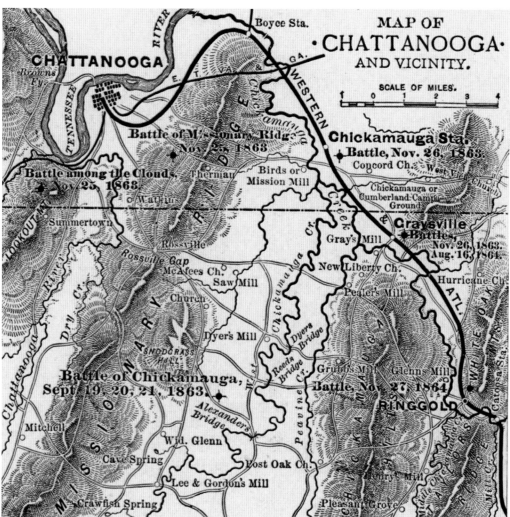

LEFT: Lookout Mountain towers above the surrounding plain to the southwest of Chattanooga. Despite it being a daunting prospect to capture it, Major-General Ulysses S. Grant successfully launched two corps under Major-General Joseph Hooker against the position on 24 November 1863.

ABOVE: The terrain around the town of Chattanooga. The town, which nestles in a bend in the Tennessee River, was totally dominated by the Confederate-held high ground of Lookout Mountain and Missionary Ridge to the southwest and southeast. Note also the location of the Battle of Chickamauga of September 1863, after which the battered Union Army of the Cumberland retreated back to Chattanooga.

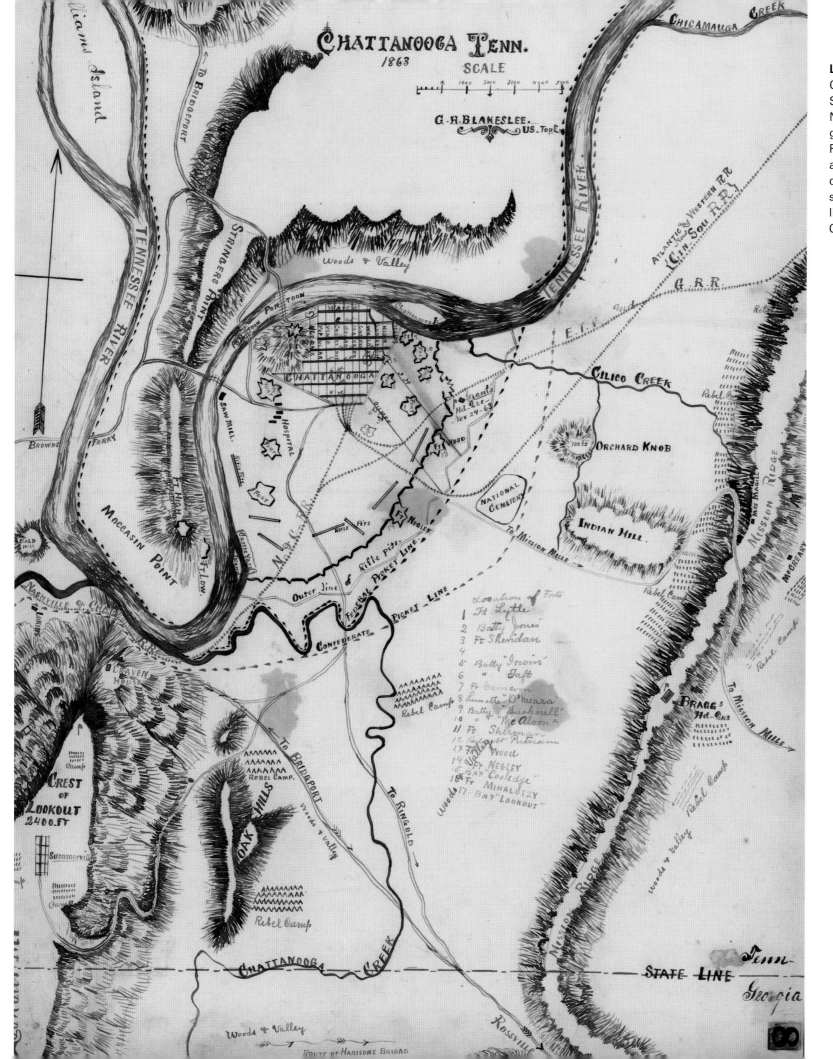

LEFT: A map of the siege of Chattanooga, which ran from late September to the last week of November 1863. This illustration, which gives little prominence to Missionary Ridge, shows the Union siege lines arcing around Chattanooga but some distance from it from the northeast to southwest, although it pays relatively little attention to those of the Confederates.

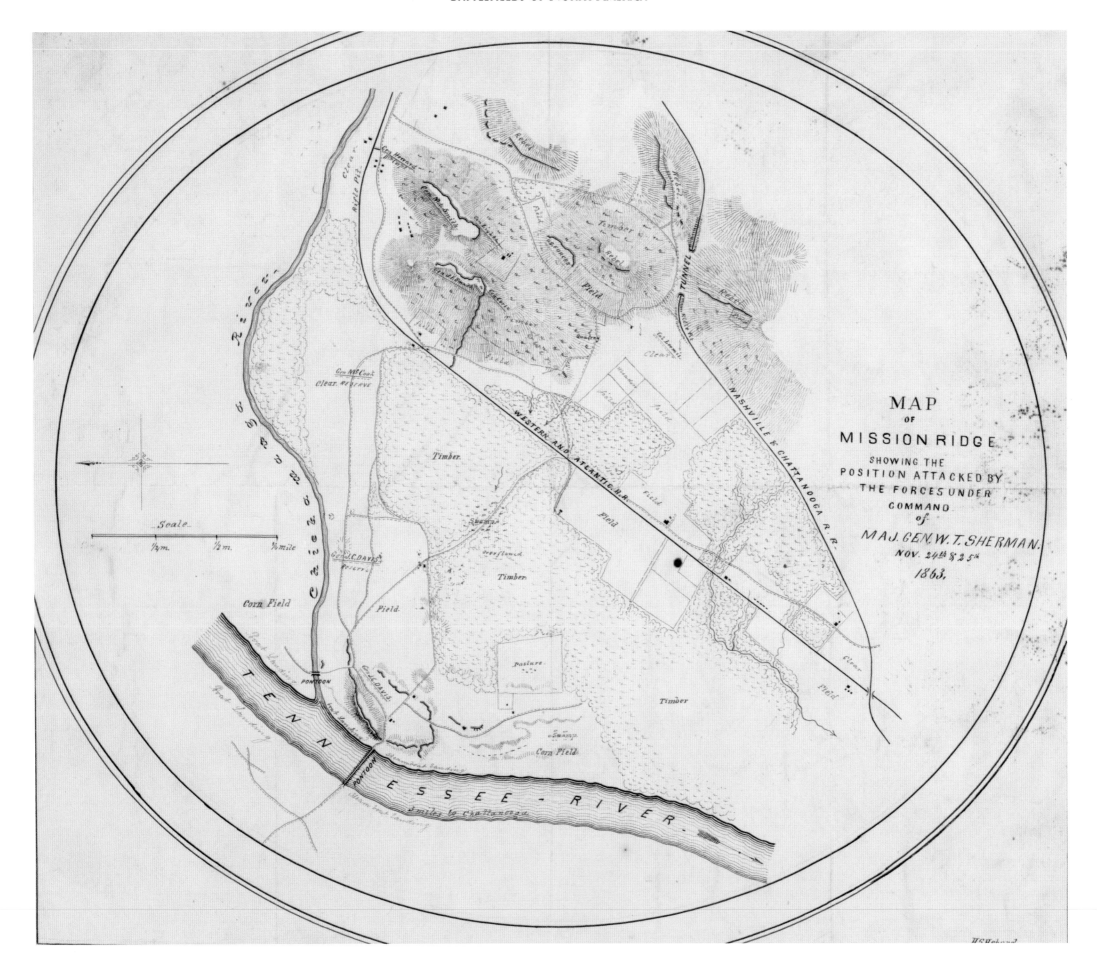

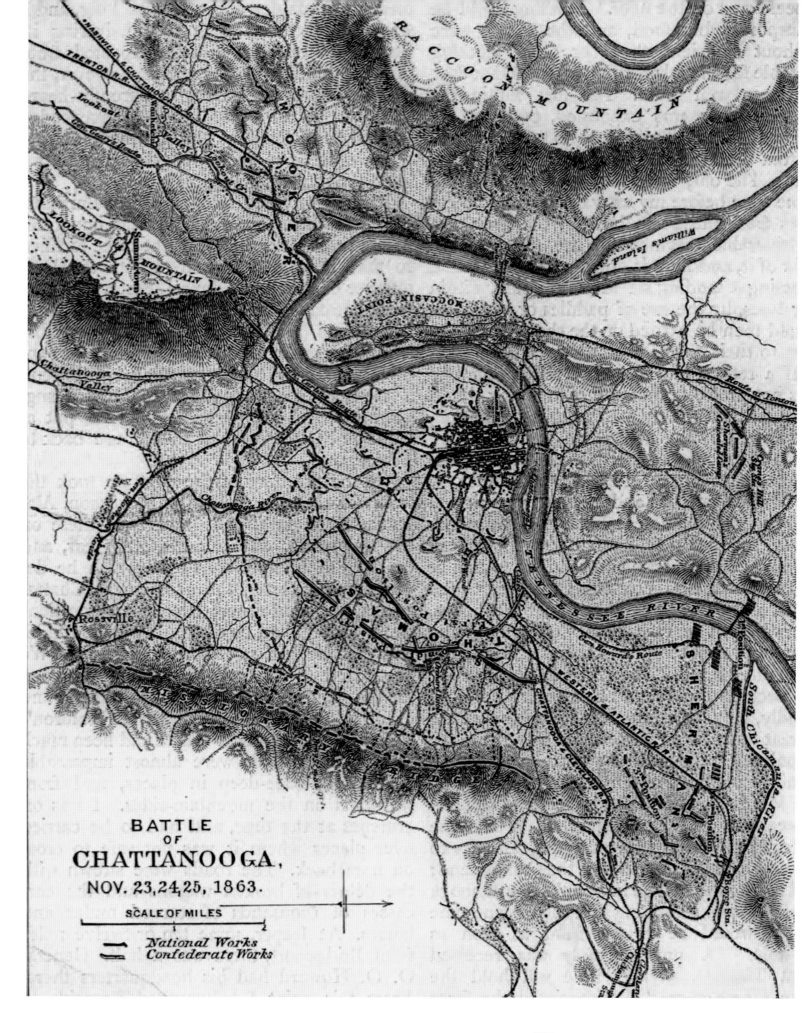

BATTLE
OF
CHATTANOOGA,

NOV. 23, 24, 25, 1863.

SCALE OF MILES

National Works
Confederate Works

FAR LEFT: Union forces under Brigadier-General William Tecumseh Sherman attacked Missionary Ridge twice during the two-day battle of Chattanooga on 24-25 November. On the first day Sherman, whose men had only just arrived in the town, flung themselves against the northern part of the ridge but were thrown back. Matters were wholly different during the second attack, which was christened the Battle of Missionary Ridge. Sherman attacked on the left and Major-General Joseph Hooker on the right, but the decisive breakthrough came in the center, where the Army of the Cumberland under Major-General George Henry Thomas cut through several Confederate defensive positions with relative ease.

LEFT: A map looking east to west over the fields of battle around Chattanooga, with more details of the rival entrenchments. Grant arrived in the town on 23 November 1863 and effectively broke the Confederate siege by putting a part of Thomas's Army of the Cumberland in pontoon boats on the Tennessee River and floating them past the besiegers on Lookout Mountain.

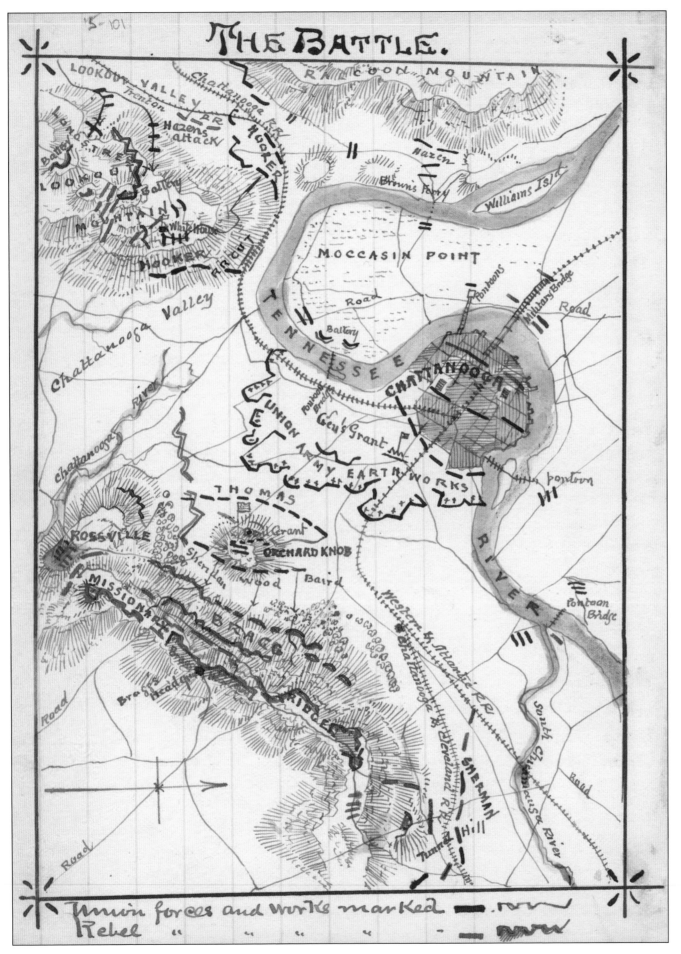

ABOVE: Brigadier-General William Tecumseh Sherman succeeded Grant as commander of the Army of the Tennessee in October 1863 and led it during the Battle of Chattanooga. Thereafter, he relieved Union troops besieged in Knoxville, a move that effectively ended the campaign and saw the Confederates abandon Tennessee completely.

RIGHT: An excellent map showing the terrain around Chattanooga and the dispositions of the various armies involved in the fighting on 24-25 November. Note also the relative positions of General Braxton Bragg's headquarters on Missionary Ridge and that of Grant on a piece of high ground beneath the ridge known as Orchard Knob.

LEFT: A detailed map of the Battle of Missionary Ridge, fought on 25 November. It shows the disposition of various Union assault forces and the entrenchments held by the Confederates. It also clearly shows the three lines of trenches that were taken in quick succession by Thomas's Army of the Cumberland in the center of the ridge. Missing, however, are the units commanded by Hooker that actually attacked around Rossville (center left).

rushed some 1,200 miles to Bridgeport, just thirty miles from Rosecrans. Matters began to improve on 17 October, when Grant was made commander of all the Union troops between the Mississippi River and Allegheny Mountains.

He moved on Chattanooga the same day, sacking Rosecrans as he went and appointing Major-General George Thomas as his replacement. Ten days later, Union troops slipped into the city, partially lifting the siege. Grant was well aware that the Confederates still held two key positions outside the city—Lookout Mountain to the west and Missionary Ridge to the south and east—and therefore called on reinforcements in the guise of Brigadier-General William Sherman's Army of the Tennessee, which was based around Memphis.

Bragg was so confident of the impregnability of his defenses that he actually sent some 20,000 men to besiege another Union army that had become stalled at Knoxville in northeast Tennessee due to supply problems. For his part, Grant felt ready to attack once Sherman's forces and the two corps of the Army of the Potomac

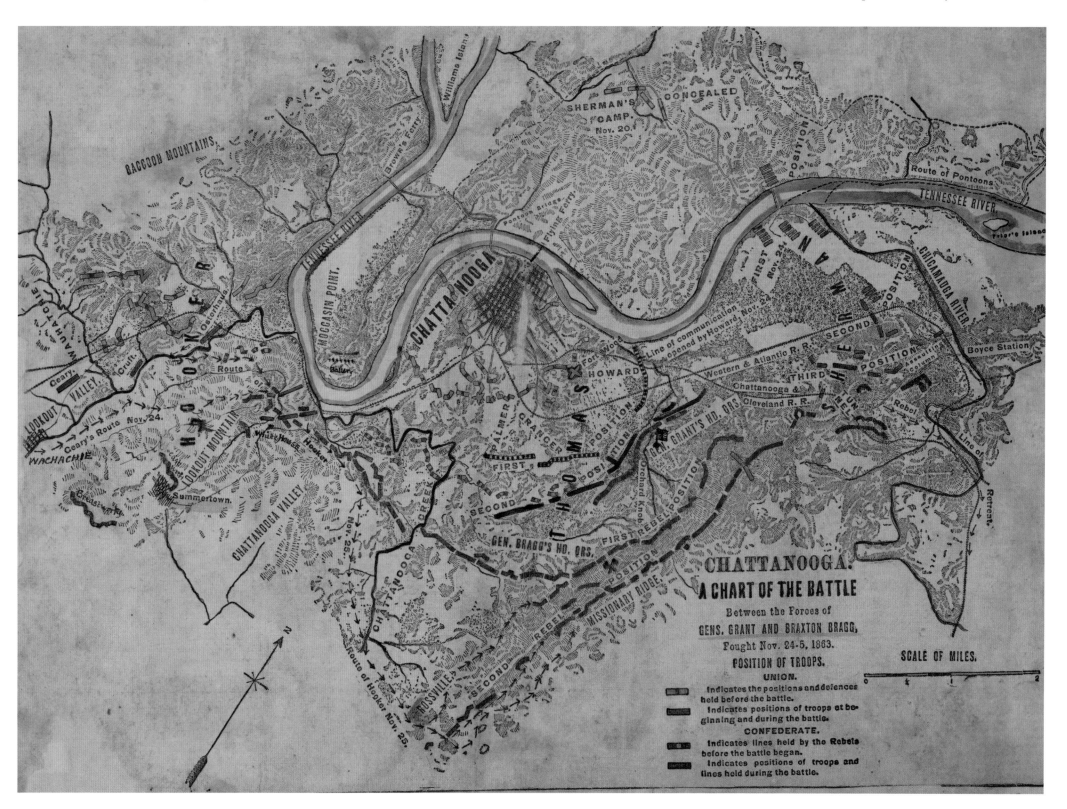

LEFT: A Currier & Ives lithograph of the Battle of Chattanooga, showing Union troops storming Confederate positions in the center of Missionary Ridge on 25 November. The attacking units of the Union's Army of the Cumberland were actually meant to have played no more than a supporting role rather than actual carrying the Confederate position. The illustration captures the spirit of the Union assault, but most of the troops wear far too pristine uniforms following several weeks of hard campaigning.

RIGHT: Major-General Joseph Hooker was something of a failure as the commander of the Army of the Potomac, a position he asked to be relieved from after his defeat at the Battle of Chancellorsville in May 1863, but he led two corps, the XI and XII, with considerable aplomb at the Battle of Lookout Mountain outside Chattanooga on 25 November, and was made a major-general of regulars (rather than volunteers) as a reward.

BELOW: The attack on Missionary Ridge as seen from the Confederate side. Waves of Union troops are advancing on the Rebel positions and the fighting appears intense. The Confederates did give way relatively quickly when the Army of the Cumberland made its unplanned assault on the center of the ridge.

had arrived. The first attack came on 24 November, when he unleashed Hooker and Sherman against Lookout Mountain and Missionary Ridge, respectively. Sherman, whose troops had only just arrived in Chattanooga, was thrown back, but Hooker's men stormed the mountain against light opposition in what was dubbed "the battle above the clouds."

The first day of the Battle of Chattanooga had captured one of the key Confederate positions, and Grant resolved to fling all of his forces against Missionary Ridge on the second day. Sherman attacked on the Confederate right and Hooker moved against the left, but neither made any significant progress. The fighting turned thanks to Thomas's Army of the Cumberland. His troops faced three lines of entrenchments but took the first line without difficulty in what was meant to be no more than a supporting attack. Rather than pause, they continued the assault, taking the two remaining lines of Confederate defenses. When they reached the top of the ridge, the defenders broke and fled. Grant then sent Sherman to relieve Knoxville in a brief campaign that ended on 6 December. Bragg was dismissed and, with Tennessee securely in Union hands, the stage was set for Sherman to launch his drive towards Atlanta, Georgia, the following spring.

THE BATTLE OF MOBILE BAY, 1864

DATE: 5 August 1864

COMMANDERS: (Union) Admiral David Farragut; (Confederate) Commodore Franklin Buchanan

NAVAL STRENGTHS: (Union) 4 ironclads and 14 other warships with 3,000 men; (Confederate) 1 ironclad and 3 gunboats with 417 men

CASUALTIES: (Union) 145 killed and 177 wounded; (Confederate) 12 killed, 20 wounded, and 123 captured

Louisiana's New Orleans, the South's most important port through which much of the supplies needed to fuel its military effort in the civil war flowed, was captured by Union forces in 1862 and thereafter Mobile, Alabama, took its place. It was not for two years that Union naval forces turned their attention to the port, but in early August 1864 Farragut, the conqueror of New Orleans, sailed into Mobile Bay at the head of a large flotilla, one that included four relatively invulnerable turreted ironclads of the famed Monitor class and fourteen more vulnerable traditional wooden warships that had been lashed together in pairs.

The smaller Confederate flotilla commanded by Buchanan was considerably less impressive, with just three wooden gunboats, *Gaines, Morgan,* and *Selma,* and a single

RIGHT: A splendid depiction of the opening of the Battle of Mobile Bay on 5 August 1864. The Union fleet is sailing from left to right and has come under intense fire from the coastal artillery within the Confederate Fort Morgan on Mobile Point. The South's only ironclad to take part is the battle, *Tennessee,* leads two of the three sidewheelers that supported it into action. Note also how the lead Union monitor ironclad, *Tecumseh,* has hit a "torpedo"—a submerged mine—with catastrophic results.

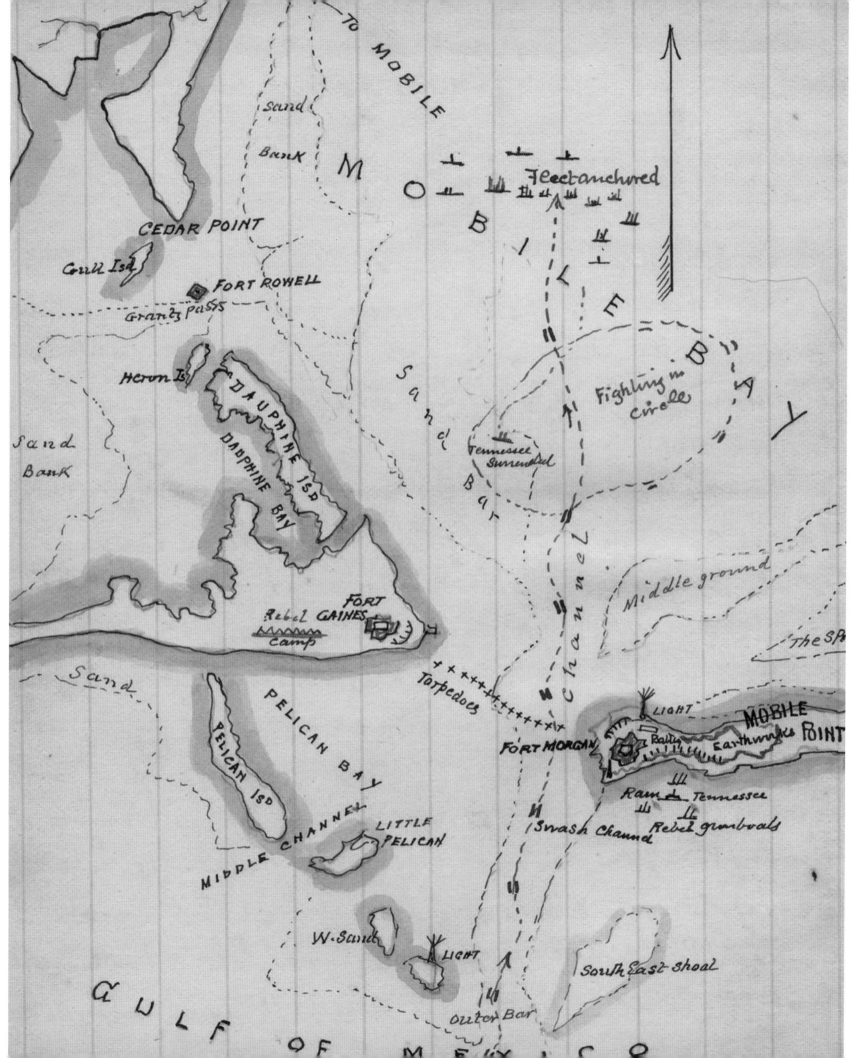

LEFT: A map showing the relative positions of the two fleets at various stages of the battle. The smaller Confederate fleet sheltered under the protective guns of Fort Morgan before the action commences (it should in fact be located on the other side of the fort within Mobile Bay), and the greater part of the victorious Union fleet eventually anchored in the bay at around 0645 hours. Note how the line of Southern "torpedoes" between the two forts considerably narrowed the entrance into the bay. The point at which the Confederate ironclad *Tennessee* surrendered is also identified; it was an event that took place at around 1000 hours, after it had been rammed into submission by a number of Union warships.

RIGHT: The Union victory at the Battle of Mobile Bay on 5 August effectively completed the blockade of the Southern states, although Forts Gaines and Powell that covered the bay were not taken until assault by Union ground forces some sixteen days after the naval engagement, and Fort Morgan held out until August. Here, Union troops raise the Stars and Stripes over the exceptionally battered remains of Fort Gaines and the adjacent lighthouse. Mobile itself remained defiant if impotent to the very end of the war.

BELOW: The interior of Fort Gaines after it had been taken by Union forces. This brickwork design was fairly typical of the coastal fortification erected to protect the US coast before the outbreak of the Civil War, whereas many of those built during the conflict were constructed of earth or sand.

RIGHT: A detailed map of the battle that shows the double line of Union warships lashed together making their way past Fort Morgan. The positions where *Tennessee's* fight ended (center top) and the sidewheeler *Selma* ran aground (top left) are also clearly marked. The Confederate ironclad was rammed repeatedly and fired on by several Union warships including *Hartford*, *Lackawanna*, *Monongahela*, and *Ossipee*, before its capitulation, while the grounded *Selma* surrendered to the sidewheel gunboat *Metacomet*.

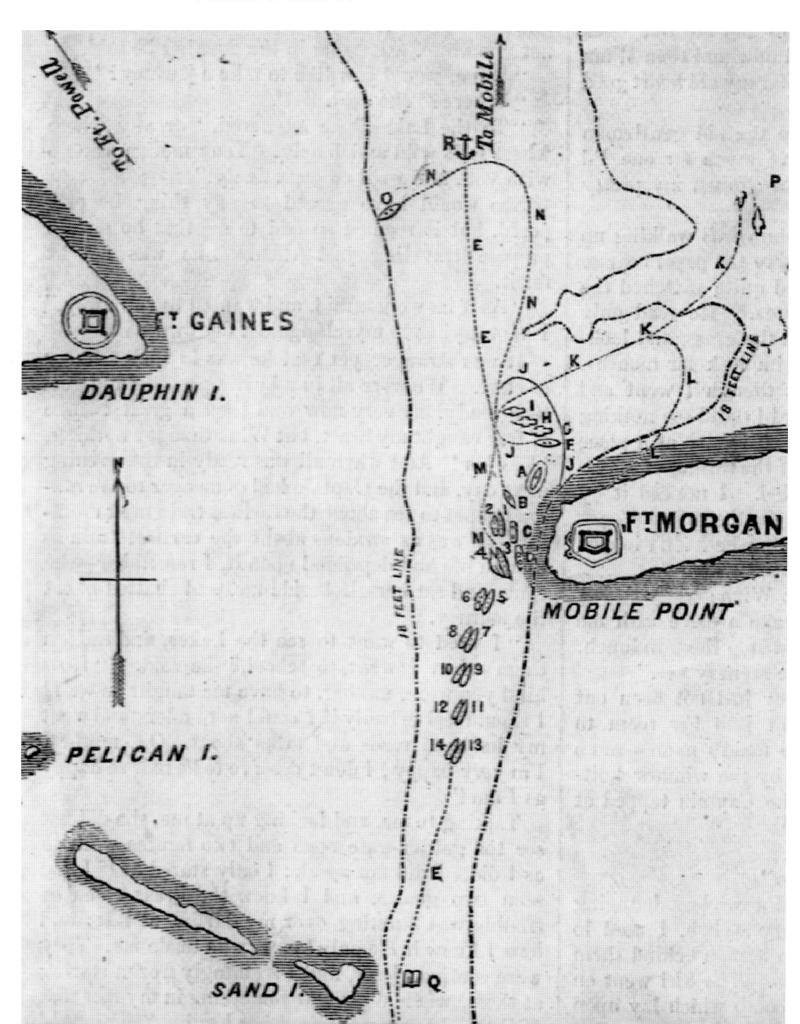

LEFT: Fort Morgan, Mobile Point, Alabama, took a terrible battering from the former Confederate *Tennessee* after the US Navy had captured and repaired the massive ironclad, and then turned its guns on the fort.

The *Tennessee* took this punishment for around sixty minutes but was itself unable to reply—it was to slow to ram the faster-moving Union warships and could not fire back because of defective fuses. The battering continued until the Southern ironclad was unable to maneuver and, with Buchanan out of action due to a broken leg, the fighting ended with his surrender at 1000 hours. Two of the gunboats were also captured. Farragut's clear-cut victory meant that the South's only really large port, and the center for blockade-running in the Gulf of Mexico, was sealed, although Mobile itself did not surrender until the last days of the war in April 1865.

ironclad, the slow-moving *Tennessee*, which was modeled on the equally renowned *Merrimac*. The action began at around 0530 hours. Farragut, on board his flagship *Hartford*, led his command into the bay but his track took him under the guns of the bay's two main forts, *Gaines* on the eastern tip of Dauphin Island, and *Morgan* at the western end of Mobile Point. The entrance is just three miles wide and the forts' fire was accurate and devastating, especially that from Morgan since it was nearest the Union flotilla.

Union casualties were severe, but much worse was soon to follow. The lead monitor, *Tecumseh*, hit a "torpedo" (actually a floating mine) as it made to attack *Tennessee* and rapidly sank. Farragut, impatient to close with the Confederate flotilla, famously ordered: "Damn the torpedoes. Full speed ahead." Buchanan moved out to attack at around 0800 hours and Farragut responded by opening fire on *Tennessee*, his most dangerous opponent, as well instructing his own *Hartford* and other warships to ram the ironclad at regular intervals.

FAR LEFT: *Hartford* was Admiral David Farragut's flagship during the battle and entered the fray lashed to *Metacomet* in an attempt to shield the less well protected smaller ship from Confederate fire. Once inside the bay they separated to improve their maneuverability in battle. *Hartford* carried both sail and a steam engine and with both in use had a speed of around 14 knots. Its overall length was 310 feet and it displaced 2,900 tons. Before going into action its upper masts and yards would have been removed.

LEFT: Admiral Franklin Buchanan (shown with an unidentified companion) commanded the Confederate ironclad CSS *Tennessee* during the Battle of Mobile Bay. As Union Admiral Farragut's counterpart, Buchanan was the Confederacy's most distinguished seaman. He had earlier commanded the CSS *Virginia*, and fought history's first battle between ironclads,

RIGHT: Short, fiery, and profane, and perhaps uninspiring to his men, Gordon Granger was one of the Union's most dependable commanders. Working closely with Farragut, he took command of the ground forces for the joint attack on Mobile.

LEFT: Farragut hangs from the tops as his flagship *Hartford* exchanges broadsides with the Rebel ironclad *Tennessee*. Having watched the USS *Tecumseh* sink after hitting a torpedo/mine, he hollered down to his fleet that was stalling, "Damn the torpedoes! Full speed ahead!"

RIGHT: Admiral David Glasgow Farragut was the top-ranking Union naval officer of the war. Before the Battle of Mobile Bay he was already a hero, having commanded his first prize ship in the War of 1812 at just twelve years of age, built up the fleet that captured New Orleans in April 1862, and cooperated in the attack on Vicksburg in 1863. At Mobile in August 1864 he steamed past the forts, ignored the torpedoes (mines), and captured or dispersed the Confederate fleet.

LEFT: Having beaten the Tennessee into submission, Farragut had the ironclad repaired and used her a few days later to force the surrender of Fort Morgan.

THE BATTLE OF THE LITTLE BIGHORN, 1876

DATE: 25 June 1876

COMMANDERS: (American) Lieutenant-Colonel George Armstrong Custer; (Native American) Crazy Horse, Sitting Bull, Gall

TROOP STRENGTHS: (American) 600; (Native American) c.2,000

CASUALTIES: (American) 53 killed and 52 wounded (Benteen and Terry), 212 (Custer); (Native American) 36 killed and 168 wounded according to Sitting Bull

US government policy in the 1870s was to force the few remaining Native American tribes living in their ancestral homelands in the Mid-West into reservations. In February 1876, the northern Sioux tribes refused to go to their assigned place in the Dakota Territory, preferring to remain in the Powder River area. Unfortunately gold was discovered in the Black Hills during 1874 and the US government tried but failed to buy them off the Sioux. The authorities then said that the Indians had to go to the reservation or face military action, thus sparking the most

LEFT: As one of their Native American scouts brings in a wounded man an isolated group of dismounted cavalrymen attempt to hold off a group of circling Sioux. The 7th Cavalry's scout detachment consisted of two officers, two interpreters, five guides, and thirty-five enlisted Native Americans. The latter were not expected to fight but some undoubtedly did.

ABOVE: Curly, one of the Sioux warriors who fought at the battle. Native Americans carried a variety of weapons, including firearms—at least thirty different types were used at the Little Bighorn—but many felt they were dishonorable weapons and preferred edged or blunt hand-to-hand weapons, bows, and lances.

ABOVE: Hunkpapa Sioux Gall acted as the main war chief during the fighting at the Little Bighorn. Gall had considerable experience of fighting the US Army as he was present at the so-called Fetterman Massacre in Wyoming during 1866, in which eighty cavalry troopers under Captain William J. Fetterman were ambushed and killed to a man, and the Wagon Box fight, also in Wyoming, the following year.

RIGHT: Lieutenant-Colonel George Armstrong Custer graduated low in his class at West Point when he graduated in 1861 but proved a dashing if self-publicizing leader of cavalry during the American Civil War. He became commander of the newly raised 7th Cavalry in 1866 and spent virtually all of the remainder of his life fighting Native Americans. His actions at Little Bighorn remain controversial, but his decision to attack without waiting for support was typical of both his dash and reckless thirst for glory.

serious clash with the Native Americans for around sixty-five years. The uprising was led by a chief of the Oglala Sioux, Crazy Horse, but partly at the instigation of Sitting Bull, a chief and medicine man of the Prairie Sioux. The various Sioux tribes were also joined by the equally disaffected Cheyenne.

The first action took place at Slim Buttes, Crazy Horse's winter home, on 17 March 1876. Some 800 mounted troops under Brigadier-General George Crook launched a surprise attack that at first scattered the Sioux, but they rallied and forced their attackers to retire. Crook was replaced in overall command by Major-General Alfred Terry, who ordered three columns to converge on the Sioux. Crook again caught up with Sitting Bull at the Rosebud River in southern Montana on 17 June and fought a bitterly contested battle, at the end of which both sides withdrew.

Terry was, at this point, out of communication with Crook and unaware of the recent fight, but he did cross Crazy Horse's trail. He therefore ordered Custer and his 7th Cavalry to get to the south of the Sioux so that they would be boxed in by the converging columns. Custer took off after the Sioux, most probably with the absolute intention of seeking battle at the earliest opportunity. On the 25th his scouts found a large Native American village on the Little Bighorn River, probably home to around 9,000 souls, but, rather than send details to his superiors or wait for Terry's and the

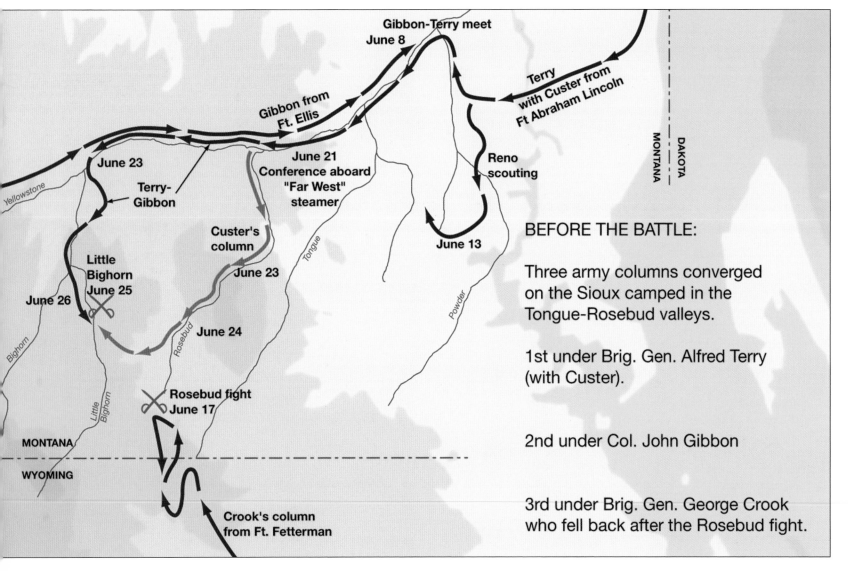

BEFORE THE BATTLE:

Three army columns converged on the Sioux camped in the Tongue-Rosebud valleys.

1st under Brig. Gen. Alfred Terry (with Custer).

2nd under Col. John Gibbon

3rd under Brig. Gen. George Crook who fell back after the Rosebud fight.

LEFT AND RIGHT: Maps showing the movement before the battle and the clash itself. In 1876 the Army's method for forcing the Sioux and their Cheyenne allies back to the agencies was to by encircle the Indians in south central Montana with three columns of troops: one commanded by Colonel John Gibbon moving east from Fort Ellis at Bozeman; the second under General Alfred Terry (with Custer) moving west from Fort Abraham Lincoln near Bismarck; and the third under General George Crook moving north from Fort Fetterman on the North Platte. The plan received a major setback when Crook's 1,300-strong column was defeated by Crazy Horse at the Rosebud fight on 17 June, forcing him to withdraw and take no further part in the campaign. Terry met Gibbon on the Yellowstone at the mouth of the Tongue on 8 June, and again on the steamer Far West near the mouth of the Rosebud. From there, on 22 June, Custer was dispatched south down the Rosebud with orders to swing northwest to the forks of the Bighorn and Little Bighorn where Terry and Gibbon would be waiting. However, Custer reached the Little Bighorn and attacked before joining forces with Terry and Gibbon.

LEFT: Brigadier-General Alfred H. Terry commanded the Dakota Column during the Little Bighorn campaign. It consisted of the 7th Cavalry, four companies drawn from the 6th and 17th Infantry Regiments, and three Gatling guns. Terry was also supported by two shallow draught sternwheel steamers, *Far West* and *Josephine*. Terry was a Civil War veteran but had not seen service since the end of that war in 1865, and thus had never fought Native Americans before.

RIGHT: Sitting Bull, the legendary Hunkpapa Sioux chief and medicine man, was a leading light in the opposition to the white gold rush into the Sioux's sacred Black Hills in the Dakota Territory during 1874. He was quickly made chief of the war council of the combined Sioux, Cheyenne, and Arapaho camp in Montana. It is often mistakenly believed that he fought at the Battle of the Little Bighorn, whereas he did not. However, the Indians who did fight maintained that he "made the medicine" that gave them their great victory. Later associated with the Ghost Dance movement, he was arrested and subsequently killed along with two of his sons by Native American police while trying to escape on 15 December 1890.

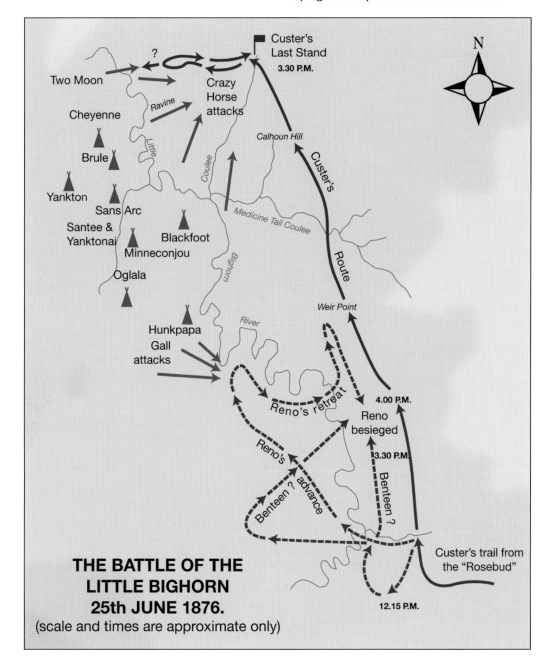

THE BATTLE OF THE LITTLE BIGHORN 25th JUNE 1876.
(scale and times are approximate only)

another column to get into position, he ordered an immediate attack.

The 7th Cavalry was split into three columns, despite it being a cardinal military sin to split a force in the face of an enemy of unknown size. One column under Captain Frederick Benteen was sent off to make sure there were no Sioux camped in a valley above the main village. Major Marcus Reno's column accompanied the one led personally by Custer. The former charged the upper end of the village but the uS troopers were thrown back across the river and pinned down. Custer made for the other end of the village but was caught in open ground before he could reach it. His

ABOVE: An early photograph of the site of Custer's last stand. This is now known as Custer Hill and this spot, on its western slope some ten to twenty yards below the crest, is where the remains of Custer, forty-one members of the 7th Cavalry and thirty-nine dead horses were found after the battle. The markers identify the points where the human remains were found. Around eighty men were cut off on the hill during the battle and part of Custer's command actually tried to fight their way out by escaping down the hill and into a ravine at its foot, but were cut down as they fled. Custer's body was found under that of a human corpse and dead horse near the crest. He had been shot in the chest and left temple.

RIGHT: Totally inaccurate misleading paintings of the Battle of the Bighorn appeared after the event. There are no high mountains near the battlefield, as suggested in this fanciful image. Furthermore, Custer's men died in groups over quite an area of ground, in gullies, ravines, and bluffs on the east side of the Little Bighorn Valley.

LEFT: Crook, one of the Native American warriors to take part in the fighting at the Little Bighorn. Although the tribes of the Mid-West plains were skilled horsemen, not all of those who fought in the battle made use of their mounts since they had developed other methods of warfare that not only relied on the mobility conferred by the horses but also of infiltration and sniping, tactics that often required them to dismount.

whole column was wiped out in a matter of an hour. Benteen was able to join Reno and they held on until relieved the next day.

The fighting continued over the next several months and reached its climax towards the end of the year. A large Sioux encampment was destroyed in the Battle of Crazy Fork Woman during a night attack on 25–26 November and Crazy Horse himself surrendered after defeat at the Battle of Wolf Mountain on 8 January 1887—yet these victories could not remove the stain of "Custer's Last Stand."

FAR LEFT: A contemporary photograph that gives a good impression of the reality of warfare on the plains of the Mid-West. The US Army tended to create mixed columns of cavalry, infantry, and artillery to hunt down the Native American tribes. Here, cavalry form a protective cordon around the column's extensive supply train and small artillery detachment. The Dakota Column, of which the 7th Cavalry was a part in 1876, had more than 150 such wagons and some 200 teamsters to manage them.

LEFT: The finishing touches to a marker commemorating a lieutenant serving with the 20th Infantry Regiment, which provided a small detachment to man the Dakota Column's Gatling Gun Battery. This consisted of three caisson-drawn guns manned by two officers and twenty-three enlisted men. Its commander was one Second Lieutenant William H. Low.

THE BATTLE OF THE BIG HOLE BASIN, 1877

DATE: 9 August 1877

COMMANDERS: (American) Colonel John Gibbon; (Native American) Chief
Joseph

TROOP STRENGTHS: (American) 206; (Native American) c.250

CASUALTIES: (American) 32 dead and 37 wounded; (Native American)
87 killed

By the late 19th Century the Native American Wars were drawing to a close, with few tribal groups remaining outside government reservations. The Nez Perce, who had never been in conflict with the white settlers, were one of the few still occupying the ancestral homelands around the Salmon River in Idaho Territory and in neighboring Oregon's Wallowa Valley. The latter area was given public-domain status in 1875 and the US Army was ordered to occupy the area. The local leader, Chief Joseph, reluctantly agreed to settle in Lapwai Reservation in northeast Idaho.

Some more hot-headed, younger members of the Nez Perce went on a drunken rampage and killed a score or so of white settlers. Joseph wanted to negotiate with the one-armed General Oliver Howard but was eventually persuaded to make for the Salmon River. A small US force under Captain David Perry set off in pursuit and on 17 June, after an arduous night march, reached the mouth of the 3,000-foot-deep canyon of White Bird Creek, where Joseph was based. When the chief's truce party was shot at, his warriors returned the fire. Perry's men soon fled.

Joseph's party now joined up with the followers of Chief Looking Glass. Howard caught up with the two chiefs on 10 July, and he opened the Battle of the Clearwater River. The Nez Perce stood and fought and suffered a number of

RIGHT: US Army troops commanded by regular army Colonel Nelson Appleton Miles arrive just in time to help Major-General Oliver Otis Howard surround Chief Joseph and his Nez Perce tribe at Eagle Creek, near Bear Paw Mountain, Montana. The "siege" began on 30 September after Miles' command suffered thirty-one men killed and forty-seven wounded in a hard-fought skirmish, but the Nez Perce were forced to surrender on 5 October.

BELOW: Chief Joseph's bid for freedom was on an epic scale as this map suggests. He set off on his prolonged march with around 800 of his people, including women and children. Around 300 of them joined Sitting Bull on the way and a further 120 died, leaving just 400 to surrender at Bear Paw Mountain. Chief Joseph, whose real name was, in translation, Thunder Rolling Over the Mountain, ended his days far from his homeland in Coleville Reservation, Washington State, where he died in 1902.

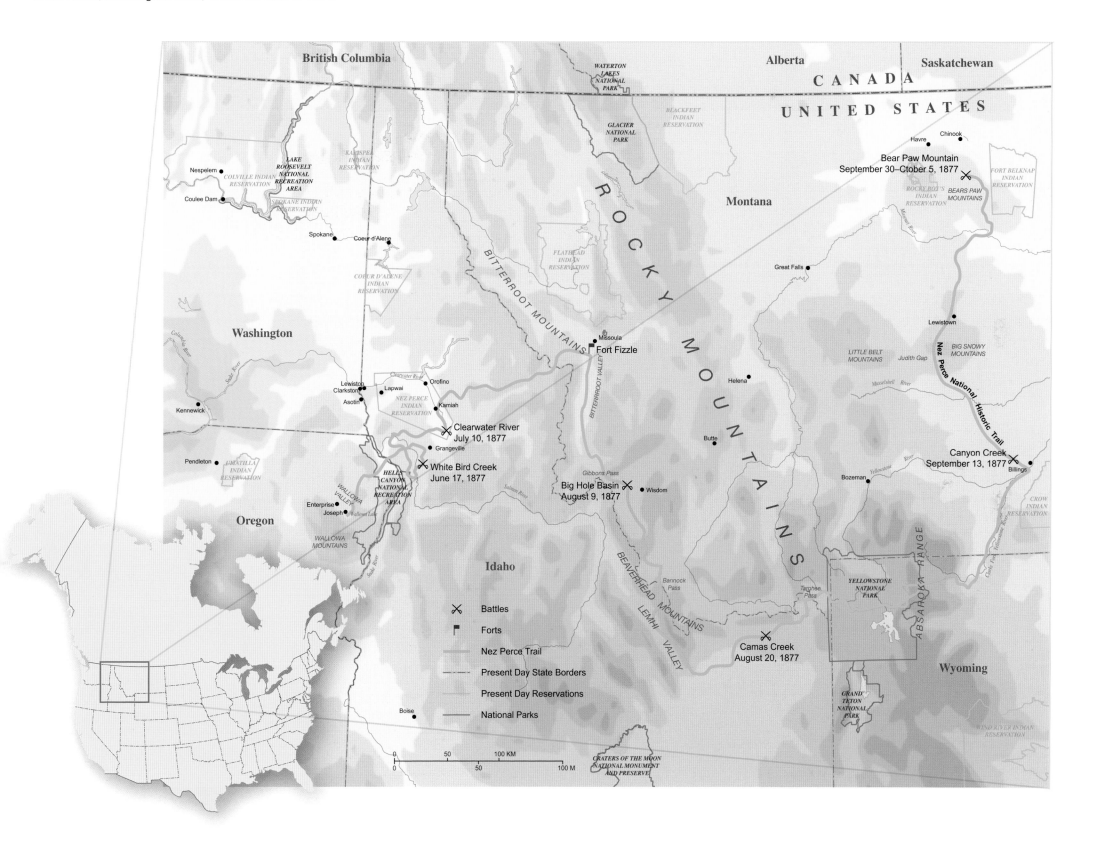

LEFT: Yellow Wolf, one of the Nez Perce warriors who followed Chief Joseph during the war of 1877. The bulk of those following the chief on the exodus toward the Canadian border were in reality women and children. Fighters like Yellow Wolf were usually in the minority.

casualties. Chief Joseph next headed into Montana Territory to hunt buffalo and if necessary make a dash for Canada. Skirting round an army post derisively known as Fort Fizzle, he opted to rest his people in the Big Hole Basin.

Howard used the telegraph to direct a force under Gibbon to intercept the Nez Perce. Gibbon's troops caught them unawares during a dawn attack in early August. Chief Joseph's camp was overrun in a mere twenty minutes but, remarkably, he was able to regroup and outfight Gibbon's men in hand-to-hand combat, pinning them down in a patch of woodland. Gibbon sent a message to Howard requesting help. The next day the Nez Perce audaciously captured their opponent's baggage train, disabled a howitzer, and captured 200 invaluable rounds of rifle ammunition. The fighting raged all day but Chief Joseph was forced to break off the action on Howard's approach. Gibbon had been comprehensively outfought.

The pursuit continued over the next several months, and Joseph turned on his pursuers at Camas Creek in Idaho Territory. Thereafter, he passed through Wyoming Territory and then turned northward back into Montana Territory. Two new US Army columns were on his trail. Joseph outwitted one at the Battle of Canyon Creek on 13 September and then halted at Bear Paw Mountain, a mere forty miles from the Canadian border. Regular Army Colonel Nelson Miles caught up with the Nez Perce on the 30th. His first attacks were repulsed, but Joseph's camp was placed under siege

ABOVE: Chief Joseph photographed in his finest ceremonial regalia. The Nez Perce leader was essentially a man of peace, one willing to compromise, but he also proved to be a more than able commander. His tactics relied on securing vital supplies, chiefly food, ammunition, and weapons, during raids or from battlefields. Using deception, surprise, and mobility, he largely outfought and outmaneuvered the US Army for some four months.

RIGHT: Civil War veteran Nelson Appleton Miles (mounted, left, in a photograph made after the Civil War) confers with one of his aides. Nicknamed "Old Bear Coat," he was a seasoned campaigner against the Native Americans, having fought in, among others, the Sioux War (1876–77). After campaigning against Chief Joseph, he fought against Geronimo (1882–86) and during the Sioux Ghost Dance campaign in 1890, which culminated in the Battle of Wounded Knee.

RIGHT: A *Harper's Weekly* engraving (based on an army officer's sketches) showing various scenes and personalities associated with the Nez Perce conflict in 1877. The US Army officers are, clockwise from top left, one-armed Major-General Oliver Otis Howard, who argued that force should be used against the Nez Perce if they could not be persuaded to move to their allocated reservation; Nelson Appleton Miles (brevetted brigadier-general of volunteers during the Civil War), whose timely arrival forced the tribe's surrender at Bear Paw Mountain; and Colonel Samuel D. Sturgis, who led six troops of the well-known 7th Cavalry Regiment during the campaign. The engraving also shows the Battle of Canyon Creek of 13 September (centre), during which Chief Joseph's Nez Perce tribe held off Sturgis and then made good their escape; a Nez Perce party capturing a stagecoach (top); and Nez Perce warriors driving ponies (bottom).

BATTLE OF CAÑON CREEK.

Camp on the Northern Pacific R.R.

Nez Percés Boy and Papoose.

Quak-um Chief of the Nez Percés.

Ft. Wal-lu-la on Columbia River.

Village of the Rovers.

Lapwai.

LEFT: A further set of engravings from the *Harper's Weekly* series depicting various places, people, and events from the campaign of 1877, including a US Army camp near the North Pacific Railroad (top left); Fort Wal-lu-la on the Columbia River (center left); a portrait of Quak-hum, a chief of the Nez Perce (upper center); and US soldiers pursuing Chief Joseph (bottom). The Nez Perce were eventually outrun by various converging US Army columns as they closed on the US-Canadian border in what was then the Montana Territory.

RIGHT: A studio portrait of Maine-born Major-General Oliver Otis Howard, who lost his right arm at the Battle of Fair Oaks in May 1861 during the American Civil War. He nevertheless went on to perform outstandingly at the Battle of Gettysburg (1863), receiving the thanks of Congress. After the war, he fought in various Native American campaigns and in 1893 finally received the Congressional Medal of Honor for his service at Fair Oaks.

for five days, and the chief was forced to seek terms on 5 October. His people were eventually exiled to Kansas and the Indian Territory, yet their long march had been remarkable.

Some 800 had covered something like 1,700 miles in three months and had suffered 120 casualties, half of whom were warriors. They had killed around 180 whites and wounded 150 or so more. They had won some battles and drawn others, and they had generally outfought, outmaneuvered, and outmarched the US Army. Miles said of them that they were "the boldest men and best marksmen of any Indians I ever encountered."

RIGHT: A further composite of events, places, and personalities involved in the Nez Perce war that eventuallycame to a conclusion in August 1877. These sketches were published in the popular magazine *Harper's Weekly*, and were based on sketches made by an army officer. The images show: top left, an Indian stronghold in ravines; upper center, George A. Huston, an army guide; above right, Nez Perce warriors in bomb-proof excavations; center left and right, sounding the bugle for the truce and sending the flag of truce to the Nez Perce camp; bottom, the advance of the skirmish line in the battle; and, lower center, the surrender on 5 October 1877 of the Nez Perce leader, Chief Joseph, to units of the US Cavalry.

THE BATTLE OF PEARL HARBOR, 1941

DATE: 7 December 1941

COMMANDERS: (American) Admiral Husband E. Kimmel; (Japanese) Vice-Admiral Chuichi Nagumo

NAVAL STRENGTHS: (US Pacific Fleet in Pearl Harbor) 70 warships and 24 support craft; (Japanese First Air Fleet) 6 aircraft carriers, 2 battleships, 2 cruisers, a destroyer screen, and 8 support ships

CASUALTIES: (American) 19 ships sunk or disabled, 164 aircraft destroyed, and 128 damaged, 2,335 service personnel killed, 68 civilians killed, and 1,178 people wounded; (Japanese) 29 aircraft, 1 large submarine, 6 midget submarines

As relations between Japan and the United States deteriorated in the late 1930s and early 1940s, the senior strategists of the Imperial Japanese Navy realized that their fleet, although modern, powerful, and large, would be hard pressed to best the equally powerful US Navy in a prolonged war. The Pacific conflict, when it came, would be dominated by these two forces and the Japanese recognized that US dockyards would out-build their dockyards in any arms race. Admiral Isoroku Yamamoto, the commander-in-chief of the Japanese navy's main strike force, the Combined Fleet, devised a plan to launch a surprise air strike against Pearl Harbor, the main base for the US Pacific Fleet on Oahu, one of the Hawaiian Islands, by carrier-borne aircraft. He hoped to cripple his opponent sufficiently to force the thus undefended United States to negotiate a peace favorable to his homeland.

Yamamoto came up with a brilliant, if complex plan in November 1941 but was not entirely optimistic, stating that even if the attack was successful it would give

LEFT: The routes taken by the first and second waves of Japanese aircraft during their final run in to Oahu. The first appeared at around 0740 hours and the second some seventy minutes later. All of the island's air bases were targeted so as to prevent US aircraft from intervening in the actual attack on the various warships at anchor in Pearl Harbor.

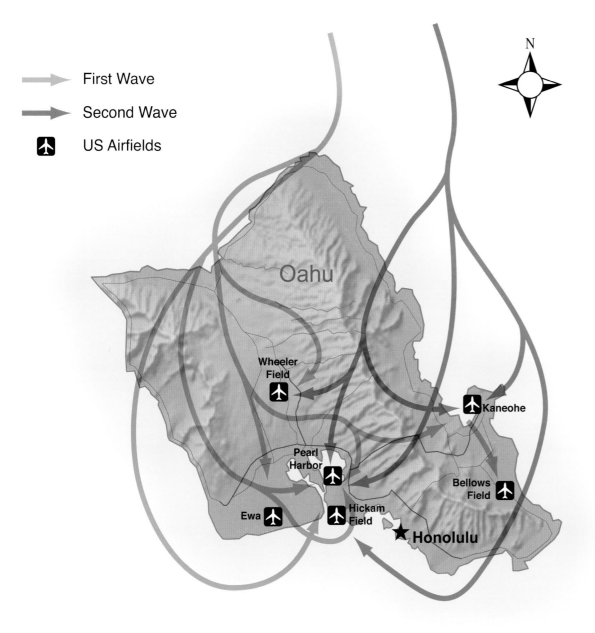

→ First Wave

→ Second Wave

✈ US Airfields

N

Oahu

Wheeler Field ✈

Kaneohe ✈

Pearl Harbor ✈

Bellows Field ✈

Ewa ✈

Hickam Field ✈

★ Honolulu

0   50 Miles   100 Miles
0   50 KM   100 KM

RIGHT: A Japanese Mitsubishi A6M2 Zero fighter takes off for the attack on Pearl Harbor as the carrier's crew looks on. This naval version of the Zero entered service in the summer of 1940 and, as with many other Japanese aircraft, stress was placed on maneuverability, range (1,940 miles with drop tank), and speed (316mph) rather than protection. Some eighty-one took part in the attack, with forty-five in the first wave and thirty-five in the second.

Japan a breathing space of only six months or so. The strike force, known as the First Air Fleet left the Kurile Islands on 25 November (Washington time) and took a circuitous route to reach a point to the north of Oahu on the 7th. Two attack waves were launched—the first of 183 aircraft and the second of 168. The first wave appeared over the islands at 0755 hours and the onslaught continued for two hours or so. The carrier aircraft targeted the island's several air bases and the vessels anchored in Pearl Harbor, especially the prestigious larger warships lying at anchor in "Battleship Row." American soldiers and sailors did fight back but the base looked a sorry sight when the Japanese departed.

Nagumo refused to allow a third strike—a decision that was a serious misjudgment.

The attack brought the United States into the war and it was soon recognized that the damage inflicted on the fleet at Pearl Harbor was not as bad as it first seemed. Most of the sunk or damaged ships were raised and repaired, the fleet's vital aircraft carriers were at sea so escaped unscathed, and the Japanese had made no attempt to knock out the base's dockside facilities or its fuel storage tanks. Admiral Kimmel did pay the price for US ill-preparedness and was soon replaced. Only six months later, just as Yamamoto had predicted, the US Pacific Fleet struck back in spectacular fashion, using its carriers to sink four of Japan's at the Battle of Midway, fought during early June.

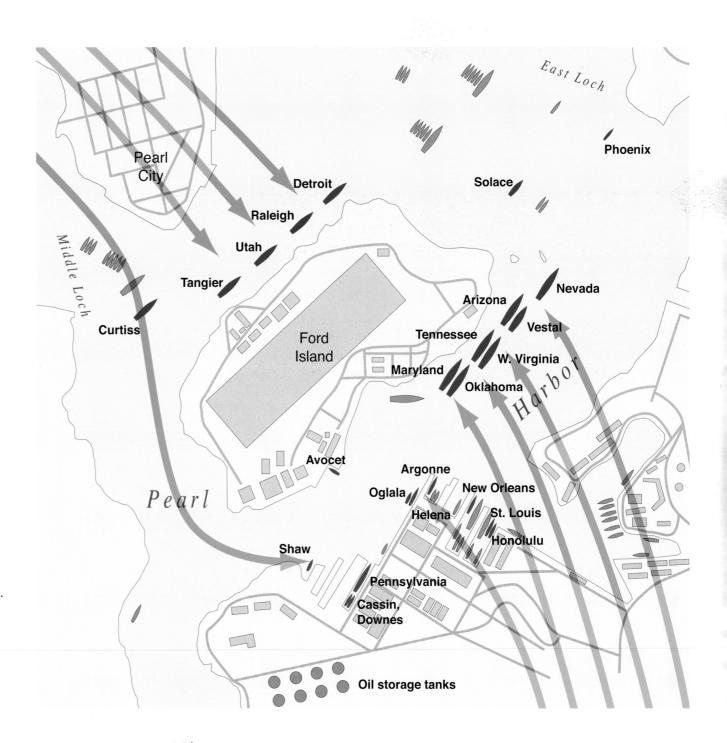

RIGHT: A map depicting the first attack on Pearl Harbor by Japanese torpedo-bombers. These were two-man Nakajima B5N2 Kate types capable of carrying a single 1,760lb torpedo. The first wave of forty Kates was led by Commander Shigeharu Murata, who served on the carrier *Akagi*, and his attack in two separate groups commenced at around 0800 hours. During the first attack the battleships *California, Oklahoma,* and *West Virginia* were hit; in he second, the cruiser *Helena* and minelayer *Ogala*; in a third, the cruiser *Raleigh* and target ship Utah.

LEFT: Vice-Admiral Chuichi Nagumo was a staunch supporter of naval aviation and commander of the Japanese First Air Fleet, which gave him effective control of the navy's six carriers at the outbreak of war and put him in immediate control of the air attack on Pearl Harbor. However, much of the plans for the attack were not in fact drawn up by Nagumo but by his chief air planner, the brilliant Minoru Genda. The vice-admiral was removed from command after losing four carriers at the Battle of Midway some six months after the raid on Pearl Harbor.

ABOVE: Admiral Isoroku Yamamoto was commander-in-chief of the Imperial Japanese Navy's Combined Fleet, its main strike force, between 1939–43 and was therefore the main architect of the country's naval strategy. He actually opposed war with the United States since he, along with several other senior officers, doubted that Japan could win such a conflict. Nevertheless, he conceived of the attack on Pearl Harbor but believed it would give his country only a breathing space and not a decisive war-winning victory.

ABOVE: The crew of the USS *California* abandon their vessel as it settles on the bottom of the comparatively shallow Pearl Harbor. The battleship was left with its quarterdecks awash but was eventually raised and repaired.

LEFT: Fire tenders and rescue boats swarm around badly damaged warships in "Battleship Row." Many in the US Navy mistakenly thought that it was not necessary to make provision to protect vessels in Pearl Harbor from torpedo attack because they believed it was too shallow for them to be used effectively. The Japanese actually devised shallow running torpedoes to overcome the problem.

ABOVE RIGHT: Smoke billows from the stricken battleship USS *West Virginia*, while the superstructure of the USS *Tennessee* can be seen in the background.

RIGHT: Brigadier-General Walter Campbell Short (seated) was a career soldier who had served in World War I and was placed in charge of the Hawaiian Department in January 1941. He was promoted to lieutenant-general the following February but was sacked after the events of 7 December and retired in February 1942.

LEFT: An aerial photograph of Pearl Harbor taken before the Japanese attack. Ford Island and its naval air station are in the center of the image, while "Battleship Row" is visible to the immediate southeast of the island; a lone carrier is berthed on its opposite side. The base was formally inaugurated as early as 1919 but there was no permanent US fleet stationed there until as late as 1940. Yet many complained that the anchorage was far from ideal. It had only one narrow entrance, was rather congested, and it took around three hours to get the entire fleet out into open water.

RIGHT: A Japanese pilot's view of the attack, with "Battleship Row" and Ford Island in the foreground and Pearl City across the water. As the smoke (which is probably from the Wheeler Field air base) suggests, other targets were hit. However, the attackers singularly failed to bomb or strafe the base's various fuel storage facilities and also missed the various workshops and engineering facilities that would be used to fully repair many of the stricken warships.

RIGHT: Admiral Husband Edmund Kimmel (center) had served in World War I and was made commander of the US Pacific Fleet at Pearl Harbor in February 1941. He was removed from his post on 17 December and replaced by Admiral Chester William Nimitz. A subsequent inquiry charged him with dereliction of duty and he retired with the rank of rear-admiral in March 1942. A Congressional board of inquiry established in 1946 eventually cleared him of the charge relating to the 1941 surprise attack.

INDEX